CAMBRIDGE LIBRARY COLLECTION

Books of enduring scholarly value

Earth Sciences

In the nineteenth century, geology emerged as a distinct academic discipline. It pointed the way towards the theory of evolution, as scientists including Gideon Mantell, Adam Sedgwick, Charles Lyell and Roderick Murchison began to use the evidence of minerals, rock formations and fossils to demonstrate that the earth was older by millions of years than the conventional, Bible-based wisdom had supposed. They argued convincingly that the climate, flora and fauna of the distant past could be deduced from geological evidence. Volcanic activity, the formation of mountains, and the action of glaciers and rivers, tides and ocean currents also became better understood. This series includes landmark publications by pioneers of the modern earth sciences, who advanced the scientific understanding of our planet and the processes by which it is constantly re-shaped.

A Stratigraphical System of Organized Fossils

William Smith (1768–1839) was a civil engineer and canal surveyor known as the 'Father of Geology' who developed and pioneered the concept of stratigraphy Through his work with canal building Smith become familiar with many different types of rock throughout Britain. He realised that fossils were specific to a certain strata of rock and that rock strata could be identified and correlated by the fossils they contained. Smith used this knowledge to publish the renowned first geological map of Britain in 1815. This volume, first published in 1817, contains Smith's catalogue of his fossil collection for the British Museum. Smith catalogues the fossils according to the rock strata and location in which they were found, together with a brief description of the fossil. This volume was the first published example of rock strata used as a cataloguing principle which demonstrated the practical nature of Smith's system of stratigraphy.

A
Stratigraphical System of Organized Fossils

With Reference to the Specimens of the Original Geological Collection in the British Museum

William Smith

CAMBRIDGE UNIVERSITY PRESS

Cambridge, New York, Melbourne, Madrid, Cape Town, Singapore,
São Paolo, Delhi, Dubai, Tokyo, Mexico City

Published in the United States of America by Cambridge University Press, New York

www.cambridge.org
Information on this title: www.cambridge.org/9781108021159

This edition first published 1817
This digitally printed version 2010

ISBN 978-1-108-02115-9 Paperback

STRATIGRAPHICAL SYSTEM

OF

ORGANIZED FOSSILS,

WITH

REFERENCE TO THE SPECIMENS

OF

THE ORIGINAL GEOLOGICAL COLLECTION

IN THE

BRITISH MUSEUM:

EXPLAINING

THEIR STATE OF PRESERVATION

AND

THEIR USE IN IDENTIFYING

THE

BRITISH STRATA.

BY

WILLIAM SMITH,

CIVIL ENGINEER AND MINERAL SURVEYOR;

Author of a "Treatise on Irrigation," a "Map and Delineation of the Strata of England and Wales," with a Memoir, "Strata identified by organized Fossils," "Geological Section" and "Table."

LONDON:

Printed for E. WILLIAMS, Bookseller to the PRINCE REGENT, and to the DUKE and DUCHESS OF YORK, No. 11, Strand, near Charing Cross; and sold by all other Booksellers in the Kingdom.

1817.

Cox and Baylis, Printers, Great Queen Street,
Lincoln's Inn-Fields.

PART I.

CONTENTS.

Indexes to Genera and Species and to their Localities will be given in the succeeding part of the work; and the Characters of the Genera will be explained.

Part II., which completes the Work, will be speedily published.

INTRODUCTION.

This novel and interesting description of near seven hundred species of Fossil Shells, Zoophites, and other organized Fossils, found in England and Wales, and collected in identification of the Strata, refers particularly to the specimens of a geological collection deposited in the British Museum. On the specimens, Roman capitals mark the genus,—the figures, 1, 2, 3, &c. refer to the species,—and the small letters, *a*, *b*, *c*, &c. to the localities or sites in the Strata.

This copious reference to the stratum which contains the Fossils, to the particular site therein whence obtained, and to the individual specimens of the collection, which is intended to be publicly exhibited in the British Museum, seemed to render figures of them unnecessary; especially as reference is constantly made to another work of the Author's now publishing by Mr. Sowerby, which consists chiefly of engravings; and as further reference is likewise made to the numerous figures of Sowerby's *Mineral Conchology*.

It is hoped, and confidently expected, that this method of combining the stratified and systematic arrangement of Fossil Shells, &c. will comprise all the information respecting them which can be required, either by the curious Conchologist or by the most attentive investigator of the Strata. Lamarck's system, which is an extension and improvement of that of Linnæus, has been preferred, as most applicable to the arrangement of organized Fossils; of which the species and even genera are sometimes not without difficulty determined. His more particular attention to the form, muscular attachment, and other striking characters apparent in Fossils where the hinge cannot often be observed, better enables the *Mineral Conchologist* to distinguish the generic and specific characters of those numerous organized Fossils which are merely casts of the inside of the shell, or of the animal which inhabited it. As this country abounds with numerous species which cannot be arranged under any of the genera given by M. Lamarck, new genera have been formed by Sowerby, Parkinson, and others, which are herein adopted and referred to: reference is likewise made to most of our old authors on the subject.

In the arrangement of Echini, I have followed *Leske* in his comment on Klein.

Four of the most copious divisions, Echini, Ammonites, Terebratula, and Zoophites, to the extent of more than a hundred species, have been arranged in a tabular form, which shows by inspection their relative situation in the Strata, and gives the effect of collating the specimens.

By the tables it will be seen which Fossils are peculiar to any Stratum, and which are repeated in others.

The order naturally suggested by the situations of the organized Fossils themselves, Zoophites, Testacea, Echini, Crustacea, Bones of large Animals, is uniformly observed in each Stratum throughout the work: so that this book naturally divided by the Strata, admits of easy reference to the innumerable specimens of this extensive branch of Natural History. In addition to these several references, to the Stratum in which any particular kind of Fossil is found, to the particular site in its course where it was collected, to the identical specimen in the British Museum, and to the Authors who have written on them, I have, for the convenience of those who have not the opportunity of visiting remote parts, given other reference to my map of the Strata of England and Wales, by colours corresponding to the coloured lines on that map, by which the courses of the Strata are represented.

My method of arranging Fossils generally, according to the Strata which contain them, having long since been adopted by all to whom my early discoveries were communicated, such connoisseurs will make by the help of this work the material improvement in their collections which is now offered to them; by which the stratified and systematic arrangements are most usefully combined. All the localities of each Species in each Stratum are enumerated in succession, showing, as the specimens would if placed together, their use in identifying the Strata.

The following works have been consulted, for the purpose of giving to the specimens the most appropriate and descriptive names:

LATIN.

Linnæus .... Systema Naturæ.
Llwyd .... Lithophylacii Brittanici Ichnographia. 8vo.
Klein with Leske ...... Echinodermata. 4to.
Brander and Solander .. Fossilia Hantoniensia. 4to.

FRENCH.

Cuvier .... Géographie Minéralogique des Environs de Paris. 4to.
Les Annales du Muséum d'Histoire Naturelle, containing Lamarck's Description of the Fossils of Grignon, Courtagon, &c. 4to.

ENGLISH.

Plot .... Natural History of Oxfordshire. fol. pl.
Morton .... Natural History of Northamptonshire. fol. pl.
Woodward............ Catalogue. 8vo. 2 vol.
Walcot .. .. Petrifactions in the neighbourhood of Bath. pl.
Ellis and Solander .... Zoophites. 4to. with Plates.
Martin .... Derbyshire Petrifactions. 4to. with Plates.
Parkinson .... Organic Remains of a former World. 4to. 3 vols. with Plates.
Townsend .... Credibility of the Mosaic History established, &c. 4to. with Plates.
Sowerby .... Mineral Conchology of Great Britain. 8vo. with Plates.

My original method of tracing the Strata by the organized Fossils imbedded therein, is thus reduced to a science not difficult to learn. Ever since the first written account of this discovery was circulated in 1799 it has been closely investigated by my scientific acquaintance in the vicinity of Bath; some of whom search the quarries of different Strata in that district with as much certainty of finding the characteristic Fossils of the respective rocks, as if they were on the shelves of their cabinets. By this new method of searching for organized Fossils with the regularity with which they are imbedded in such a variety of Strata, many new species have been discovered. The Geologist is thus enabled to fix the locality of those previously found; to direct the attentive investigator in his pursuits; and to find in all former cabinets and catalogues numerous proofs of accuracy in this mode of identifying the Strata.

The virtuoso will therefore now enter upon the study and selection of organized Fossils with the twofold advantage of amusement and utility. The various component parts of the soil, and all the subterraneous productions of his estate, become interesting objects of research; the contents of quarries, pits, wells, and other excavations, hitherto thought unworthy of notice, will be scrupulously examined.

The organized Fossils which may be found, will enable him to identify the Strata of his own estate with those of others: thus his lands may be drained with more certainty of success, his buildings substantially improved, and his private and public roads better made, and repaired at less expense.

To possess such general knowledge of the Soil, Subsoil, and Strata, on every road he travels, and in every field he traverses, with their respective Fossils stratigraphically arranged in his cabinet, must furnish an endless source of gratification to every inquisitive mind. His own house will be the best school of Natural History for all the younger branches of the family, and a source of amusement to all his scientific acquaintance;—science will become more general;—all will lend assistance;—no time will be idly lost: nor with such resources can a country gentleman be (as Pope says) " a prisoner in his own house every rainy day." Rural amusements, to those who can enjoy them, are the most healthful; and the search for a Fossil may be considered at least as rational as the pursuit of a hare,—one the sport of infants, the other of adults; one squanders time and property, the other improves the mind and may afterwards extend such infant knowledge to the improvement of the estates he may enjoy. No study, like that of the organized Fossils, can be so well calculated for the healthful and rational amusement of youth; and nothing can more effectually lead to general improvement than the early diffusion of science. That of Natural History should be the first object of every country gentleman; and if it be not an insult to nature to pass unnoticed her various productions, which are superior far to the nicest workmanship, it will sometime be an insult to the understanding to be considered totally ignorant of these things.

The Author is aware that for want of proper and settled terms there may be some mistakes in his arrangement of the Fossils; but if a *Venus* should be mistaken for a *Cyclas*, or a *Mactra* for a *Mya*, it cannot be of more importance (in the present state of the science) than the misspelling of a word which alters not the sense; —errors of *misnomer* in the work or in the tables may be easily corrected with a pen, as more perfect specimens are obtained, or as the science improves, but errors in their stratified arrangement can be corrected by those only who are locally acquainted with the Strata, and the numerous organized Fossils they contain. On this principle I have ventured, without much knowledge of Conchology, and with weak aids in that science, to give the outlines of a systematic arrangement combined with the stratified, it being much easier to learn the useful than to unlearn the useless; and finding myself pressed to the task by others, who are proceeding with such works with but imperfect knowledge of the Strata, I may therefore hope that my imperfections in the systematic arrangement placed against their's in the Strata, the balance will be in favour of this work.

A combination of the stratified with the systematic arrangement will always have the advantage of showing the locality of specimens, and from the simplicity of form, errors in genera and species may be readily corrected, and the species hereafter discovered as readily added.

My observations on this and on other branches of the subject are entirely original, and unincumbered with theories, for I have none to support: nor do I refer my reader to foreign countries to prove what I advance: but I have drawn and described the face of a country whose internal contents are more deeply explored than any other part of the earth's surface; and in which every one, to the extent of his local knowledge, is a critic on my work. With such a host of critics as might rise up against me in the extensive districts I describe, it may reasonably be supposed that accuracy claimed my utmost attention, and that in the collection and arrangement of the numerous Fossils, and in the determination of the facts which they support, the mind would frequently hesitate, and doubts have remained some time for want of specimens or an opportunity of visiting the site of any dubious part. Rules, however, have arisen out of the research, which wonderfully assist the investigation of Strata; these will be described in the course of my works, if the encouragement which I expect from the public should enable me to publish all my manuscript. The extent and accuracy of the observations which I have made and recorded will excuse me for any apparent procrastination, and encouragement only is wanting in this important branch of natural History to unravel the mystery and simplify our knowledge of all the terrene part of the creation; for enough is already discovered to render the subject familiar and extensively useful. Geologists are also aiding greatly an improvement in the science by their minute examination of the materials of Strata, and by the modern adoption of a settled plan of observing and recording natural facts.

This particular branch of Geology has already proved that a large portion of the earth once teemed with animation, and that the animals and plants thus finely preserved in the solid parts of the earth's interior, are so materially different from those now in existence, that they may be considered as a new creation, or rather as an undiscovered part of an older creation. They are chiefly submarine, and as they vary generally from the present inhabitants of the sea, so at separate periods of the earth's formation they vary as much from each other; insomuch that each layer of these fossil organized bodies must be considered as a separate creation; or how could the earth be formed *stratum super stratum*, and each abundantly stored with a different race of animals and plants. Surely these innumerable and finely organized fossils are not the sports of nature placed there to excite the attention of the idly curious, but they must, like the other works of the Great Creator, have their use. The miner has long thus identified his local sinkings of shafts, and why should not the virtuoso, the owner and occupier of the soil, quarry-men and workers of stone, do the same more extensively? The result of my labours is a settled plan for doing this, and therefore the identification of Strata by the help of organized Fossils, becomes one of the most important modern discoveries in Geology. It enables the Geologist clearly to distinguish one Stratum from another in Britain, and also to trace their connexion with the same Strata on the continent. Thus it is capable of the most extensive or of the most local use.

Strata, when identified by organized Fossils, with the localities of each Stratum accurately delineated, will also define on maps of the country the limits of districts in which they are most prevalent. Thus it appears by this work, and by the map of the Strata, that three principal families of organized Fossils occupy nearly three equal portions of the English and Welsh parts of Britain;

Echini are most common to the superior Strata;

Ammonites to those beneath;

Producti with numerous Encrini to the inferior; or at least to the lowest of those which are most abundantly stored with the animal class.

This order in the section gives the following order on the map;

Echini most abundant in the eastern and south-eastern districts;

Ammonites in the interior, or in the course of those Strata which occupy the midland counties in a north-easterly direction, from Charmouth to Whitby;

The species of Anomia. *Linn.* now called Productus, occur no higher in the series than in the limestones which are the boundaries of coal.

Taking another extensive view of a singular division of the animal class, it appears by the peculiar localities of certain species, and even several genera of these organized fossils peculiar to certain Strata, that many of them occur not throughout large districts of the country. Thus, the Corals, which accompany the Pisolite and the two Oolites in the greatest abundance, are not found within the *denudation*, a district which is composed of large portions of Kent, Surry, Sussex, and Hampshire; and more locally known as the Forest Ridge and the Wealds of Kent and Sussex. Two similar expansions of the same Strata occasion a similar deficiency of the numerous species of these extensive genera, on North York Moors, and on Blackdown, which embraces a considerable portion of the counties of Somerset, Dorset, and Devon.

So far the deficiency arises from a want of exposure in those Strata which contain them; but another cause of the deficiency arises from defects in the courses of those Strata: two of the three rocks before mentioned are not discovered north of the Humber, or south west of the borders of Somerset, and the other is but partially seen to the south west of these limits; consequently a still larger portion of Yorkshire in the north east and of the parts beforementioned in the south west (wherein these fossils might have been expected) is deficient in them.

These deficiencies arise from the *unconformableness* of some of the other Strata.

Other deficiencies of organized Fossils, to a considerable extent, on many parts of the surface where they should, according to the course of their out-cropping strata, be found, are occasioned by a thick covering of alluvial matter; this, in the rounded fragments or nodules of the different Strata, contains entire many of the specimens herein described, or loose ones which have been detached from their original beds in the Strata, much rounded by attrition in water: these will be described under the head of *Alluvial Fossils.*

Fortunately for the science which these investigations must establish, many Strata are entirely deficient in organized Fossils, some of these so thick as to occupy of themselves large portions of the island, and some much thinner, which alternate with others that contain them.

How far these facts tend to establish a certain theory, which pretends to give the relative ages of Strata according to the presence or absence of organized Fossils, others may determine.

Many Strata being entirely without organized Fossils, the investigation is much facilitated, by rendering the courses of those Strata which contain them more distinct; and the courses of all the Strata being known, the name of the place where any specimen is found is sufficient to mark its locality in the Strata, and the specimens being filled with the matter in which they are imbedded materially assist in identifying the Stratum to which they belong. In this respect Mineral Conchology has much the advantage of recent; the matter of the Stratum fully compensating in a geological point of view, for any defect in the specimen. Shells are generally without the animals, which are mostly incapable of preservation; fossils frequently represent the animals without the shells (*i. e.* the interior conformation of the shell). In general, fossil shells are so effectually closed and filled with stony matter, that the hinge, opening, and other characters, cannot be observed.

Numerous Zoophites naturally too tender for preservation, have in their fossil state their shape and most minute organization beautifully retained in limestone, flint, and other solid matter. Thus not only in clays, sands, and rocks, but in the hardest stones, are displayed all the treasures of an ancient deep, which prove the high antiquity and watery origin of the earth; for nothing can more plainly than the Zoophites evince the once fine fluidity of the stony matter in which they are enveloped, no fluid grosser than water being capable of pervading their pores. The process which converted them and their element into stone seems to have been similar to that of freezing water, which would suddenly fix all the inhabitants of the ocean, each in its place, with all the original form and character. Organized Fossils are to the naturalist as coins to the antiquary; they are the antiquities of the earth; and very distinctly show its gradual regular formation, with the various changes of inhabitants in the watery element.

Thus endless gratification may be derived from mountains of ancient animated nature, wherein extinct animals and plants innumerable, with characters and habits distinctly preserved, have transmitted to eternity their own history, and the clearest and best evidence of the earth's formation.

As the section of a tree shows its increase by annual rings of growth, so the Strata seem to show the earth's lamellar increase.

The interior of the earth therefore, like other better investigated works of nature, is formed upon the wisest and best principles, incomprehensible in their extent, but to the limits of our capacity exceedingly useful.

The angle of declination in the Strata towards the horizon, and their consequent successive termination in the surface of the earth, is one of the most wonderful dispositions of Providence that could be devised for the benefit of man. Without such an angle of the Strata with the surface of the earth, coals and other useful minerals could not be worked; nor could the edges of the Strata have been thus identified by the organized Fossils imbedded. They are therefore of great use in tracing the outcrops of open-jointed Strata, which alternate with clays, and produce springs; as also in identifying the Strata of deep wells sunk to those subterraneous reservoirs in the cavities of rocks: and the vegetable impressions particularly define in the collier's shaft the approach to coal. They are found in their respective Strata at the greatest depths, as well as at the greatest heights. Hence the consideration of such an immensity of animal and vegetable matter, the time required for its perfection and subsequent consolidation in the Strata, evidently in deep and quiet water, may at first seem incomprehensible to many, but further investigation leads to further admiration of the great unerring cause. In this, as in every other part of the creation, there seems to have been one grand line of succession, a wonderful series of organization successively proceeding in the same train towards perfection.

Zoophites and shell fish appear to have been the first of the animal creation, and to have been in existence prior to the completion of dry land, as they have entered into the composition of a large proportion of the solid parts of the earth. Many Strata of considerable extent are little else but an accumulation of their remains, which have been quietly entombed in the Strata, with all the form, characters, and habits of life, in the places where they are; there being no appearance of their having been disturbed or drifted about. The change of the watery element in which they lived, from a fluid to a solid state, has been the cause of their destruction and the means of their fine preservation; and that change seems in most instances to have been produced without violence: the form of the tenderest substance is as finely preserved in the hardest stone as in the original state.

These numberless appearances of an animated origin in Strata of such immense thickness and extent, must strike the admirers of nature with a degree of reverential awe and grateful adoration of the Almighty Creator. And though the mysterious cause which placed them there comes not within our comprehension, these remains of animal and vegetable life are the clearest evidence of the boundless extent of creation; that one destroying and reanimating power has ever existed, and that nothing is done in vain. If these animals and vegetables had only to live and die, and mark respectively the sites of their existence in the mass of matter which now forms the earth, they have had their use, and will for ever remain indefaceable monuments of that wonderful creative power which formed them and all things.

The position in which these organized Fossils are arranged seems to be the most useful part of the knowledge to be derived from them; is it not therefore better to profit by that which is in our power, than to waste our time about that which is incomprehensible? not that there is reason to fear any result from man's deepest investigation of the works of nature, which must end in the acknowledgement of an omnipotent, incomprehensible, allwise, directing agent.

The chief object of this work being to show the utility of organized Fossils in identifying the Strata, nothing further will be attempted in the systematic arrangement than is necessary to make the subject intelligible; and the numerous useful and interesting deductions thence resulting will more appropriately follow than precede the regular description of them in the order of the strata.

The term "Organized Fossils" *is generally applied to all fossil matter that has a relation to the form of any organized body, either animal or vegetable.*

These substances are also called "*Fossils*," "*Petrifactions*," and "*Organic Remains*."

GEOLOGICAL TABLE OF BRI

WHICH IDENTIFY THE COURSES AND CONTINUITY OF

AS ORIGINALLY DISCOVERED BY W. SMIT

GEOLOGICAL MAP OF

ORGANIZED FOSSILS which Identify the respective STRATA.		*NAMES OF STRATA on the Shelves of the* GEOLOGICAL COLLECTION	COLOURS *on the* MAP of STRATA	*NAMES in the MEMO*
Volutæ, Rostellariæ, Fusus, Cerithia, Nautili, Teredo, Crabs Teeth, and Bones	*Plains*	*London Clay*		London Clay *forming Highg*
				Clay or Brickearth *with Inte*
Murices, Turbo, Pectunculus, Cardia, Venus, Ostreæ		*Sand* *Crag* *Sand*		Sand & light Loam *upon a*
Flint, Alcyonia, Ostreæ, Echini ... *Plagiostoma*	*Chalk Hills*	*Chalk* *Upper*		Chalk *Upper part soft con*
Terebratulæ, Teeth, Palates ... *Plagiostoma*		*Lower*		*Lower part hard an*
Funnelform Alcyonia, Venus, Chama, Pectines, Terebratulæ, Echini		*Green Sand*		Green Sand *parallel to the*
Belemnites, Ammonites	*Clay Vales*	*Brickearth*		Blue Marl
Turritella, Ammonites, Trigoniæ, Pecten, Wood		*Sand* *Portland Rock*		Purbeck Stone Kentish Rag Pickering *and* Aylesbury
Trochus, Nautilus, Ammonites in Masses; Ostreæ (in a bed) Bones		*Sand* *Oaktree Clay*		
Various Madreporæ, Melaniæ, Ostreæ, Echini, and Spines		*Coral Rag and Pisolite*		Iron Sand & Carstone *whic* *Fuller's Earth and i*
Belemnites, Ammonites, Ostreæ		*Sand* *Clunch Clay and Shale*		Dark blue Shale *producing*
Ammonites, Ostreæ		*Kelleways Stone*		*in North W*
Modiola, Cardia, Ostreæ, Avicula, Terebratulæ	*Stonebrash Hills*	*Cornbrash*		Cornbrash *A thin Rock of*
		Sand & Sandstone		
Pectines, Teeth and Bones, Wood		*Forest Marble*		Forest Marble Rock *thin Bed*
Pear Encrinus, Terebratulæ, Ostreæ		*Clay over the Upper Oolite*		Great Oolite Rock *which pr*
Madreporæ		*Upper Oolite*		
Modiolæ, Cardia		*Fuller's Earth & Rock*		
Madreporæ, Trochi, Nautilus, Ammonites, Pecten		*Under Oolite*		Under Oolite *of the Vicinity*
Ammonites, Belemnites as in the under Oolite		*Sand*		
Numerous Ammonites		*Marlstone*		
Belemnites, Ammonites in mass	*Marl Vales*	*Blue Marl*		Blue Marl *under the best*
Pentacrini, Numerous Ammonites, Plagiostoma, Ostreæ, Bones		*Lias*		Blue Lias White Lias
		Red Marl		Red Marl and Gypsum *sot*
Madreporæ, Encrini in Masses, Producti		*Redland Limestone*		Magnesian Limestone Soft Sandstone
Numerous Vegetables, Ferns lying over the Coal	*Coal tract*	*Coal Measures*		Coal Districts *and the Rock* *generally*
Madreporæ, Encrini in Masses, Producti Trilobites	*Mountainous*	*Mountain Limestone*		Derbyshire Limestone *or M*
		Red Rhab & Dunstone		Red & Dunstone *of the Sou* *Interspersion*
		Killas		*Various* Killas or Slate *and other Str* *West Side of* *of Limestone*
		Granite, Sienite & Gneiss		Granite Sienite and Gneiss

From the reexamination of the Authors numerous Specimens in the arrangement of his Geological Collection in the British Museum and his subsequent observ

A Geological Map of England and Wales in 15 Sheets, dedicated by permission

…TISH ORGANIZED FOSSILS,

…THE STRATA *IN THEIR ORDER OF SUPERPOSITION;*

…TH, *Civil Engineer;* WITH REFERENCE TO HIS

…*ENGLAND AND WALES.*

…IR and the PECULIARITIES of the STRATA.	PRODUCTS of the STRATA.
…gate, Harrow, Shooters and other detached Hills	Septarium from which Parker's Roman Cement is made
…rspersions of Sand and Gravel	No Building Stone in all this extensive District but Abundance of Materials which make the best Bricks and Tiles in the Island
…Sandy or absorbent Substratum	Potters Clay, Glass Grinders Sand, and Loam and Sands used for Various Purposes
…tains flints	Flints the best Road Materials
…tains none	Good Lime for Water Cements
…Chalk	Firestone and other soft Stone sometimes used for Building
…g and Limestone of the Vales of	The first Quarry and building Stone downward in the Series Kimmeridge Coal
…s in Surry and Bedfordshire contain …n some Places Ochre and Glass Sand	Fullers Earth, Ochre and Glass Sand Some Lime used on these Sands in Sussex and Yorkshire
…n strong Clay Soil chiefly in Pasture …lts and Vale of Bedford	
…Limestone chiefly arable lying in Clay	Makes tolerable Roads
…s used for rough Paving and Slating	Coarse Marble, rough Paving and Slate
…oduces the Bath Freestone … of Bath and the midland Counties	The finest Building Stone in the Island for Gothic and other Architecture which requires nice Workmanship
…Pastures of the midland Counties	Excellent Lime for Water Cements Now used for printing from M.S. written on the Stone
…t Sandstones and Salt Rocks and Springs	
	Small Quantities of Copper and Lead and Calamine
…s & Clays which accompany the Coal … a Sandstone beneath	Grindstones, Millstones, Pavingstone, Iron-Stone and Fire-Clay from the Coal Districts
…Metalliferous Limestone	Lead, Copper, Calamine Marble
…thern and Northern Parts with …s of Limestone marked blue	Some good Building Stone
…ata of the Mountains on the …f the Island with Interspersions … marked blue	The Limestone polished for Marble Tin, Copper, Lead and other Minerals The most durable building Stone in the Island for Bridges and other heavy Works

Numerous Trials for Coal.

Part on which Lime is rarely used as a Manure

Greatest extent of good land.

Mines and Mineral Districts.

Part on which Lime is generally used

…tions this list of the Strata has been improved and his future exertions will be in proportion to the encouragement which he receives from the Public.

…to S.r Jos.h Banks, P.R.S. by W. Smith, also by the same Author, Strata identified by organized Fossils.

LONDON CLAY.

TESTACEA.

UNIVALVIA.

VOLUTA.

Voluta spinosa. *Sowerby's Mineral Conchology.*

Fig. 5 and 6, London Clay Plate, Strata identified, &c.

A cast of the inside of the shell; oblong, with a short spire and not many volutions, a few longitudinal undulations on the upper part of each.

Bognor.

FUSUS.

Fusus longævus. *Min. Conch.*

Ventricose, smooth, spire turreted with a few large knobs upon the upper part of the latter whorls; beak as long as the spire, slightly curved near the end.

Muddiford.

CERITHIUM.

SPECIES 1.

Cerithium melanioides. *Min. Conch.*

Fig. 7, London Clay Plate.

Turreted, volutions convex, with one spiral largely tuberculated carina above the middle, and several tuberculated carinæ beneath it; obscurely undulated longitudinally: beak very short.

a Woolwich.

b Bracklesham bay.

SPECIES 2.

Cerithium intermedium. *Min. Conch.*

Pyramidal, sides straight: volutions many, the upper edge distinctly crenated, and several small crenulated threads beneath; a concavity at the meeting of the volutions: mouth squarish.

a Woolwich.

b near May Place.

SPECIES 3.

Cerithium funatum. *Min. Conch.*

Conical, elongated, with two obtuse crenulated transverse ridges upon each whorl; upper part of each whorl thickened and tuberculated; mouth squarish; base smooth, with two transverse ridges.

Newhaven Castle Hill.

This species has fewer volutions than species 2, the row of tubercles on the upper edge is larger, nor has it so many spiral threads beneath.

VIVIPARA.

Vivipara fluviorum. *Min. Conch.*

Fig. 1, *London Clay Plate.*

Volutions four to six, convex, shell about twice the length of the aperture; lines of growth rather sharply conspicuous, giving the shell a finely striated appearance.

Brixton Causeway, out of a deep well.

MULTILOCULAR or CHAMBERED UNIVALVES.

NAUTILUS.

Nautilus imperialis. *Min. Conch.*

Involute, umbilicate; aperture lunate; septa entire, concave, broadest in the middle, truncated and slightly recurved at their ends; siphunculus nearest to the inside.

Isle of Sheppey, a small specimen.

AMMONITES.

Ammonites communis. *Min. Conch.*

Fig. 11, *LondonClay Plate.*

Involute; volutions six or more, exposed; radii numerous, prominent, bifurcating over the front; aperture nearly circular, equal to about one-fifth of the diameter of the shell.

Happisburgh cliff.

There are two varieties of this species, one of them has a flattish spire, with an oval aperture; the other a rounder spire, with a nearly circular aperture.

This specimen is of the first variety.

BIVALVIA.

EQUIVALVED BIVALVES.

MODIOLA.

Modiola depressa. *Min. Conch.*

Much depressed, ovate, narrowing toward the posterior side; surface smooth.

Bognor.

PECTUNCULUS.

SPECIES 1.

Pectunculus decussatus. *Min. Conch.*

Fig. 10, *London Clay Plate.*

Transversely obovate; sides rather straight; surface covered with numerous longitudinal striæ: hinge teeth twenty-five to thirty: margin thick, plain.

Highgate archway.

SPECIES 2.

Pectunculus.

Fig. 3, *London Clay Plate.*

Orbicular, rather depressed, thickest near the beaks; longitudinally undulated; undulations indistinct, alternately wider and narrower: many concentric rugæ.

Bognor.

CYCLAS.

SPECIES 1.

Cyclas deperdita, *Lam.*

Ovato-triangular, depressed, beaks acute, anterior side subtruncated; surface with sharp transverse lines of growth.

Woolwich.

The middle tooth is frequently bifid, and the lateral ones crenulated.

SPECIES 2.

Suborbicular or transverse, beaks acute, the anterior side straightest; surface with acute transverse elevations.

a Woolwich.

b Newhaven Castle Hill.

Differs from the last species in the more central position of the beaks.

VENERICARDIA.

SPECIES 1.

Venericardia planicosta. *Min. Conch.*

Subcordate, very thick, smooth; ribs broad and flat, about twenty, expanding into each other toward the margin; a few large teeth within the posterior edge.

Bracklesham Bay.

SPECIES 2.

Subcordate, oblique, beaks lying toward the less produced side; many longitudinal flattened undulations indenting the margin, and crossed by transverse rugæ; inner edge of the shell furnished with undulations equal in number to those on the outside :—shell thick.

Bognor.

This species resembles venericardia senilis, but the undulations are much flatter and more numerous.

TELLINA.

Fig. 2, *London Clay Plate.*

Ovate, wider than long; transversely striated.

a Sheppey.

b Happisburgh cliff.

MYA.

Mya intermedia. *Min. Conch.*

Depressed, smooth, twice as wide as long, sides rounded; anterior side expanded, gaping a little; posterior side small; front nearly straight.

Bognor.

INEQUIVALVED BIVALVES.

CHAMA.

Chama squamosa *Brander.*

Fig. 4, *London Clay Plate.*

Subrotund, valves convex, with transverse squamous furrows; beak of the lower valve most projecting.

Hordel cliff.

OSTREA.

Oblong, a little oblique, rather smooth; lower valve gibbous in the middle, depressed at the margin; pit oblong, narrow.

Woolwich.

CRUSTACEA.

CRAB.

Several species of Crabs have been found at Sheppey, but I have not the opportunity of distinguishing them.

BONES.

TEETH.

Fig. 8, *London Clay Plate.*

Large, conical, one side flat, smooth; with two fangs or small teeth, one on each side.

Isle of Sheppey.

Fig. 9, *London Clay Plate.*

Thin or much depressed, the point oblique, with two fangs or small teeth, one on each side.

Happisburgh cliff.

CRAG.

ZOOPHITA.

Spheroidal or depressed, with concentric cavities or cells; surface with numerous angular depressions, covered by the points of very small aggregated tubes; no stem appears.

Aldborough.

Orbicular; with numerous fasciculi of small tubes radiating from the stem; the upper surface covered by the aggregated points of tubes.

Aldborough.

FLUSTRA.

Surface undulated; with minute openings placed on small elevations, radiating from the root or place of attachment in a quincuncial order.

Aldborough.

TESTACEA.

UNIVALVIA.

EMARGINULA.

Emarginula reticulata. *Min. Conch.*

Fig. 5, Crag Plate, Strata identified, &c.

Shell oval, reticulato-striated, vertex rather acute, principal radii twenty-four or more.

Bramerton.

VOLUTA.

Voluta Lamberti. *Min. Conch.*

Fusiform, short, smooth; base elongated, obscurely truncated; columella with three or four plaits; aperture acute above; outer lip sharp, not expanded toward the base.

Aldborough—cast of the inside.

MUREX.

SPECIES 1.

Murex striatus. *Min. Conch.*

Shell ventricose, with many spiral rounded projections, and from three to five parallel striæ between each, crossed by longitudinal sutures: volutions from three to six; beak nearly straight; mouth oval.

a Alderton.

c Bramerton.

SPECIES 2.

Fig 2, Crag Plate.

Oblong or ovate; volutions five or six, ventricose, with many spiral threads, (alternately larger and smaller); longitudinal sutures distinct; aperture oblong, wide in its upper part, ending below in a short deep canal; the right lip grooved at the edge.

These grooves correspond in number and in size to the threads on the outside, the lip generally acute. In most of the specimens, two of the larger threads in the upper part of the volution are more conspicuous than the others, and continue to the apex; the upper one of the two forms a slight angle in the contour; the alternation of larger and smaller threads is not always regular.

a Between Norwich and Yarmouth.

b In the parish of Leiston.

c Thorpe Common.

d Tattingstone Park.

e Bramerton. A specimen from this place is nearly covered with small Balani.

Foxhole. One from this place is covered with larger Balani.

Playford.

Sutton. These specimens have the upper larger thread very distinct.

A VARIETY approaching Murex latus. *Min. Conch.*

a Trimingsby.

b Bramerton.

c Thorpe Common.

SPECIES 3.

Oblong, obtuse, volutions about five, with eight or more spiral ridges, two of which in the upper part are larger than the others, and are continued to the apex: longitudinal sutures distinct; aperture oblong, straightened on the inner side, wide in the upper part; canal short, broad.

Of the two projecting ridges the upper is the largest, and there is a considerable hollow between them, with two or more parallel striæ. A small ridge on the upper edge of the volution.

a Bramerton.

b Thorpe Common.

Sutton.

This species may be distinguished from the others by its large and few ridges.

SPECIES 4.

Murex contrarius. *Linn. Min. Conch.*

Fig. 1, *Crag Plate.*

Spire reversed, volutions five or six, slightly expanded at the upper part, and contracted toward the beak: surface either smooth or with many rounded projections: mouth irregularly ovate; beak rather short.

a Alderton.
b Suffolk.
c Tattingstone Park.
Foxhole.
Playford.
Sutton.
Newborn.
Brightwell.

SPECIES 5.

Murex rugosus.? *Min. Conch.*

Subturreted or oblong; volutions rather flattened, and suddenly contracted into a short beak, with several large longitudinal undulations crossing many deepish spiral striæ.

Foxhole.

The inside of the shell is undulated in the same manner as the outside.

SPECIES 6.

Murex antiquus. *Linn.*

Ovate, wide at the base; volutions about seven, expanded, with many spiral projections, which are crossed by longitudinal undulations in the upper part: aperture oblong, right lip sharp, with internal undulations corresponding with the external projections.

Surface reticulated; aperture expanded at the base; left lip spread.

Playford.

TROCHUS.

SPECIES 1.

Conical; volutions five or six, the lower part projecting, with many finely crenulated lines; aperture rhomboidal, transverse; no umbilicus; base convex.

Trimingsby.

SPECIES 2.

Volutions four or five, the middle expanded into a keel or ridge, which continues to the apex; many spiral striæ; aperture half the length of the shell, rounded on its left side, angular on the right; no umbilicus.

Bramerton.

TURBO.

SPECIES I.

Turbo littoreus. *Min. Conch.*

Fig. 3, *Crag Plate.*

Shell suboval, acute, striated; columnar margin flat: volutions about five.

a Between Norwich and Yarmouth.

b Leiston, Old Abbey.

c Thorpe Common.

d Bramerton.

VARIETY.

Subglobose; volutions four or five, flat or concave at their upper edge, striated; aperture equal to more than half the length of the shell.

a Bramerton.

b Thorpe Common.

SPECIES 2.

Obtuse; volutions four or five, with one spiral projection not continued to the apex; striated; aperture half the length of the shell.

Bramerton.

SPECIES 3.

Obtuse; volutions four or five, with two spiral projections, the upper one continued to the apex; striated.

Bramerton.

TURRITELLA.

SPECIES 1.

Fig. 4, *Crag Plate.*

Pyramidal, sides straight; with three continued crenated lines; the middle one smallest.

Thorpe Common.

SPECIES 2.

Turritella trilineata.

Volutions flattish, lower part angular, sometimes projecting, with three smooth spiral threads and obscure spiral striæ: base with several spiral striæ.

Length 1 to 2 inches, volutions about ten.

The lower part of the volution angular, with a projecting edge and a small thread above it; upper part thinned, a slight thread beneath it: a projecting thread on the middle. Base rather convex.

a Bramerton.

b Trimingsby.

Foxhole.

Sutton.

SCALARIA.

Scalaria similis. *Min. Conch.*

Whorls contiguous, spire with five or six rounded transverse elevations, close to each other, and somewhat decussated, the lowest most prominent: ribs distant, circular.

Bramerton.

AMPULLARIA.

Subglobose, spire short, depressed; volutions four, smooth, with a canal at the upper edge; aperture oblong, a little lunate on the inner side; umbilicus often covered by the inner lip.

Trimingsby.

NATICA.

Natica glaucinoides. *Min. Conch.*

Nearly globose, spire rather elongated; umbilicus simple, partly covered; upper part of each whorl slightly depressed.

Trimingsby.

BELEMNITES.

A part of a long slender one; alluvial.

Alderton.

The core of a belemnite with many septa: flint, transparent; alluvial.

Aldborough.

SERPULA.

Serpula crassa. *Min. Conch.*

Shell acutely conical, round within, three-sided externally, four or five times as long as the diameter of the end at the aperture.

Aldborough.

BIVALVIA.

EQUIVALVED BIVALVES.

MYTILUS.

Oval or oblong; one side straight; smooth, with transverse lines of growth; hinge with three or four small teeth at the beak; shell thin.

Brightwell.

PECTUNCULUS.

Pectunculus glycimeris.

Arca glycimeris. *Linn.*

Fig. 7, *Crag Plate.*

Orbicular, depressed; surface nearly smooth, with many obscure longitudinal striæ, and concentric lines of growth; inner edge sharply dentated, fifteen or more teeth on each side of the arched hinge, about ten of which are large and each bent at an angle; shell strong.

a Thorpe Common.

b Tattingstone Park.

Foxhole.

Aldborough.

Sutton.
Newborn.
Brightwell.
Bentley.

CARDIUM.

Fig. 8, *Crag Plate.*

Obovate, wider than long, rather oblique, one side straightish; about twenty-five longitudinal ridges and furrows, crossed by transverse risings: sulcated on the inside.

a Tattingstone Park.
b Happisburgh Cliff.
c Bramerton.
Foxhole.
Sutton.
Newborn.
Brightwell.

The longitudinal ridges are rather wider than the furrows; thes hell is often thinner than that of C. edule.

It has some resemblance to C. Parkinsoni, *M. C. Tab.* 49, but is wider and has seldom more than twenty-five ribs; the inside appears to be more sulcated.

MACTRA. *Linn.*

Ovate, transverse, depressed, nearly equilateral; sides straightened, front margin straightish; smooth, with distinct lines of growth.

a Between Norwich and Yarmouth.
b Trimingsby.
c Tattingstone Park.
d Bramerton.
Sutton.
Brightwell.

VENUS.

SPECIES 1.

Venus equalis. *Min. Conch.*

Uniformly convex, obcordate or nearly circular, covered with numerous transverse concentric striæ; thick, particularly in the middle; margin acute, extended, entire: cicatrix obscure.

Newborn.
Foxhole.
Minsmere Iron Sluice.

SPECIES 2.

Ovate, anterior side straightened; with transverse rugæ, and obscure longitudinal striæ; an ovate obscure lunula on the posterior side; shell thickened at the edge.

Newborn.

ASTARTE. *Min. Conch.*

Ovate, wider than long, depressed; beak small, pointed, with a narrow flattish space on each side of it; lineated transversely; margin denticulated on the inside.

The anterior space long, narrow; the posterior ovate.

Var. 1. Rather circular; the anterior flat space narrow; beak almost central.

Bentley.

Var. 2. Beak almost central, and much flattened; anterior and posterior spaces large.

Newborn.

Var. 3. Ovate, beak not flattened.

Newborn.

Sutton.

VENERICARDIA.

Venericardia senilis.

Cordate, oblique, rather depressed, beak acute; surface with about twenty longitudinal large undulations crossed by many transverse rugæ: inner edge largely dentated; shell thick.

Sutton.

Newborn.

Bentley.

TELLINA.

Tellina bimaculata. *Foss. Hanton. Fig.* 102.

Subrotund, or ovate, wider than long, depressed, smooth, with transverse lines of growth; anterior side straightish, posterior rounded; fold of the shell slight; beak acute.

a Bramerton.

b Between Norwich and Yarmouth.

c Trimingsby.

Brightwell.

MYA.

Mya lata. *Min. Conch.*

Fig. 9, *Crag Plate.*

Ovate, depressed; anterior side acuminated and truncated, slightly gaping.

Aldborough, a cast of the inside.

Trimingsby, hinges.

Bramerton, hinges.

INEQUIVALVED BIVALVES.

OSTREA.

SPECIES 1.

Nearly orbicular, depressed, with obscure longitudinal ridges and many transverse squamæ; pit not large, oblong, shallow; shell smooth on the inside.

It is thinner than the shell of ostrea edulis: and is most concave near one side, the edge of which is reflected.

Sutton.

Bentley.

Newborn.

SPECIES 2.

Oval, oblong, depressed, one side hollowed; rough with imbricated lines of growth; pit large, oblique: shell strong.

A large irregular rough shell; one side is often hollowed.

Damerham.

PLACUNA.

Orbicular, flat; inside smooth, with a wide outer space; teeth short, diverging from the beak, and connected with the outer space; a small tubercle above the angle formed by the meeting of the teeth.

The outer surface appears to be striated and perhaps echinated; the shell is very thin. The teeth diverge at an obtuse angle, and their flat surfaces are toward each other.

Aldborough.

PECTEN.

Orbicular; depressed, two-eared; lower valve convex, with twenty or more longitudinal undulations, composed of longitudinal striæ crossed by numerous acute sutures, which give it an echinated appearance: upper valve concave, with flat risings, striated and crossed by acute sutures.

Aldborough.

Newborn.

Sutton.

TEREBRATULA.

Terebratula spondylodes.

Oval, rather depressed, with sharp transverse lines of growth; a large circular foramen in the beak; two projecting thick teeth in the lower valve: shell thin, depressed on each side of the beak.

Foxhole.

Newborn.

Aldborough, an upper valve. ?

The large perforation in the beak is grooved circularly, and also the recess beneath the beak, shell thin, except at the teeth of the hinge.

MULTIVALVIA.

BALANUS.

Balanus tessellatus. *Min. Conch.*

Fig. 6, *Crag Plate.*

Obliquely conical, thin; valves six, obscurely ribbed, smooth; interstices finely tessellated; aperture oval.

a Aldborough.

b Bramerton, on a murex.

c Burgh Castle.

d Tattingstone Park.

e Keswick.

Foxhole, on a murex.

The REMAINS of LARGE UNKNOWN ANIMALS, as teeth and vertebræ, with teeth and vertebræ of many smaller species apparently marine, occur more frequently on the eastern shore and along the course of this than in any other stratum; but alluvial matter being common to large portions of the districts which produce them, and their sites in the strata being not well defined, particulars of these reliquia will be given at the end of the series with the general remarks on the alluvial Fossils.

CHALK.

ZOOPHITA.

ALCYONIUM.

SPECIES 1.

Fig. 2, *Upper Chalk Plate, Strata identified, &c.*

Conical or pearshaped, with large openings on the sides.

The openings are often irregularly heartshaped, or with one rather acute angle and two projecting rounded ones; they, however, vary into other forms even in the same specimen; the sides of the openings appear striated.

Chittern.

SPECIES 2.

Fig. 1, *Upper Chalk Plate.*

Long, conical, with concentric undulations, and many unequal lateral openings; top very concave.

Wighton.

Near Warminster.

Guildford.

SPECIES 3.

Spreading, with small lateral openings placed in quincuncial order.

Upton.

SPECIES 4.

Long, conical, with longitudinal furrows.

a Knook.

b Chittern.

SPECIES 5.

Oblong, clubshaped, top striated, spreading.

Warminster.

Several species of Alcyonia, minutely organized, in hydrophanous flints.

Wilts.

ENCRINUS.

A scale of the tortoise encrinite.

Near Warminster.

TESTACEA.

UNIVALVIA.

TROCHUS.

SPECIES 1.

Volutions four or five, depressed, angular ; base rounded or swelling : length more than an inch, breadth twice the length : umbilicus large.

Heytesbury.

SPECIES 2.

Volutions five or six, angular, the upper part flat, the side flattish or slightly depressed in the middle ; base depressed, spirally furrowed ; aperture wide, umbilicus small.

a Heytesbury.

b Mazen Hill.

SPECIES 3.

Fig. 8, *Lower Chalk Plate.*

Depressed, volutions four or five, the upper part flat ; a projecting rim on each margin of the side, the lower one rounded beneath.

Mazen Hill.

CIRRUS.

Fig. 5, *Lower Chalk Plate.*

Short conical, volutions about five, round, smooth ; umbilicus not large.

Length more than an inch.

Near Warminster.

TURBO.

SPECIES 1.

Conoidal or subturreted, volutions five or six, the upper part with two sharp carinæ, beneath which are three or four lesser ones ; mouth round, grooved at the edge.

The spiral carinæ were probably edged with prominences on the shell, although they do not appear so on the cast of the inside.

Heytesbury.

SPECIES 2.

Conoidal, volutions five or six, round, even.

Near Warminster.

TUBULAR IRREGULAR UNIVALVES.

SERPULA.

Fig. 3, Upper Chalk Plate.

Cylindrical, nearly straight, smooth, with distant annular risings; opening round.

Norwich.

MULTILOCULAR UNIVALVES.

BELEMNITES.

Fusiform, apex papillary, an obscure furrow descending from it; flattened on one of the sides.

Norwich.

Slender, elongated.

Hunstanton cliff.

AMMONITES.

Ammonites tuberculatus.

Fig. 4, Lower Chalk Plate.

Involute, keeled; inner volutions half exposed, surface rough with large tubercular projections; tubercles oblong on the inner margin, furcating into two subnodose ridges; back wide, edged by large pointed knobs: keel entire.

The ridges proceeding from the knobs on the inner margin are each elevated into a smaller tubercle on the middle of the volution, and then are either obscurely united into one knob on the back, or otherwise form separate knobs: aperture oblong, about two-fifths of the largest diameter, which is sometimes three inches.

Norton.

BIVALVIA.

INEQUIVALVED BIVALVES.

PLAGIOSTOMA.

Plagiostoma spinosa. *Min. Conch.*

Obovate, longitudinally furrowed, sides nearly equal, straightish; one valve spinous, spines on some half the length of the shell.

The flatter valve is spinous, the number and length of the spines not constant. The angular indentation between the projecting beaks is formed in the spinous valve. Inside of the shell furrowed.

a Heytesbury.

b Near Warminster.

Guildford.

OSTREA.

Fig. 5 *and* 6, *Upper Chalk Plate.*

Nearly circular, smooth; lower valve deep, sublobate, attached; upper valve convex externally at the beak, reflexed in the front; margins wide, crenated within near the hinge; shell thin; muscular impression roundish.

Pit of the hinge small, shallow; the upper valve sometimes irregularly striated from the beak. The lower valve often projects at the back in the manner of the Gryphites, with a narrow hinge and less oblique pit; attached sometimes by nearly the whole surface, the unattached part being in the front.

Norwich.

PECTEN.

SPECIES I.

Fig. 8, *Upper Chalk Plate.*

Nearly flat, smooth; front semicircular; a few concentric flat risings, and obscure longitudinal striæ; ears two, rectangular, striated.

Norwich.

SPECIES 2.

Oblong, very depressed; valves nearly equal, with broad longitudinal ribs, and alternating furrows on which are placed one or two elevated ridges.

Mazen Hill.

TEREBRATULA.

SMOOTH OR NOT PLICATED.

SPECIES 1.

Terebratula carnea. *Min. Conch.*

Depressed, smooth, obtusely five sided, front edge short, entire; valves equally convex, slightly flattening along the middle.

Front margin not undulated; sometimes a small part of the lateral margin near the beak is reflexed.

Norwich.

SPECIES 2.

Terebratula subundata. *Min. Conch.*

Nearly circular, depressed, smooth, valves equally gibbous; front margin straight or slightly depressed in the middle, with one undulation on each side.

a Heytesbury.

b Near Warminster.

c Mazen Hill.

Guildford.

Var.—The under valve regularly arched, the upper considerably depressed; oval, middle of the shell flattish.

Fig. 9, *Upper Chalk Plate.*

Norwich.

SPECIES 3.

Terebratula semiglobosa. *Min. Conch.*

Nearly circular, gibbous, smooth; largest valve deepest and uniformly gibbous, front margin undulated, with two risings in the smaller valve.

a Mazen Hill.

b Heytesbury.

c Near Warminster.

PLICATED.

SPECIES 4.

Fig. 7, *Lower Chalk Plate.*

Rather gibbous, valves almost equally convex, front margin elevated in one large wave rounding into the sides; plicæ numerous, entire; beak acute, projecting.

The lateral plicæ are blended with those on the middle; width less than $\frac{3}{4}$ of an inch.

a Near Warminster.

b Heytesbury.

c Mazen Hill.

d Norwich.

SPECIES 5.

Obscurely five sided, striated; upper valve most convex, the front sharply elevated in an angular wave with about seven distinct short plaits; sides with twelve or more distinct plaits on their edges; beak projecting, slightly incurved.

From each side of the beak of the lower valve runs a sharp angle, connecting with the lateral margin: elevation of the front sudden, with a straight upper edge, and straight subparallel sides: the plaits on the edges of the valves short, and dividing into fine striæ which continue to the beak. The plaits vary much in number.

a Norwich.

b Heytesbury.

INOCERAMUS.

SPECIES 1.

Inoceramus Cuvieri. *Sowerby.*

Fig. 1, *Lower Chalk Plate.*

Heytesbury.

Knook Castle and Barrow.
Bury St. Edmonds.
Hunstanton Cliff.

SPECIES 2.

Fig. 2, *Lower Chalk Plate.*
Shell thin.
Near Warminster.
Guildford.

MULTIVALVIA.

BALANUS.

Fig. 4, *Upper Chalk Plate.*
Valves long, conical, with transverse furrows, one side turned inwards.
Norwich.

ECHINI.

ANOCYSTI.

CIDARIS.

SPECIES 1.

Circular, depressed, rays biporous; twenty rows of articulated eminences, which are nearly equal on the base but unequal above, the two rows in each areola diminishing faster than those in the areæ; edges of the mouth turned inwards. Diameter about an inch.

Northfleet.

SPECIES 2.

Circular, depressed; ten rows of articulated eminences, two rows in each area. Diameter about an inch.

Wilts.

SPECIES 3.

Circular, depressed; shell thin, rather smooth.?
Cidaris coronalis. *Klein.?*
Wilts.

PLEUROCYSTI.

SPATANGUS.

RAYS PLACED IN FURROWS.

SPECIES 1.

Oblong, cordate, rather high, margin rounded; dorsal furrow narrow, the ray contained in it not distinct; dorsal ridge rising higher than the apex: covered with granulæ, particularly on the middle of the base.

Wilts.

SPECIES 2.

Cordate, depressed, dorsal furrow narrow; covered with granulæ, particularly on the middle of the base.

a Norwich.
b Wilts.

SPECIES 3.

Cordate, margin rounded, dorsal furrow large, ridge high; height of the shell two-thirds of the length.

a Lexham.
b Chittern.
c Wilts.
d Smitham Bottom.
e Near May Place
f Great Ridge.
g Surry.
h Chesterford.
i Bubdown.
k Pewsey.
l Norwich.
Guildford.

RAYS NOT PLACED IN FURROWS.

SPECIES 4.

Long-cordate, the top rounded, with a short deep groove to the mouth; rays obsolete.

Wilts.

SPECIES 5.

Cordate, base flat, top convex, dorsal furrow slight, rounded; rays ten biporous lines diverging in pairs, pores large.

a Upton. A mass.
b Chittern.

CATOCYSTI.

CONULUS.

SPECIES 1.

E. Albogalerus. *Leske.*
a Wilts.
b Burnham Overy.

SPECIES 2.

E. Vulgaris. *Leske.*
Wilts.

SPECIES 3.

Pentagonal, depressed; rays of pores ten, biporous, flexuous, depressed; the area swelling into rounded angles: mouth and vent rather large.
Lexham.

GALEA.

SPECIES 1.

Fig. 10, *Upper Chalk Plate.*
E. Ovatus. *Leske.*
a Norwich.
b Taverham.
c Croydon.
Guildford.
Bury.

SPECIES 2.

E. Pustulosus. *Leske.*
a Moushold.
b Taverham.
c Lexham.
d Carshalton.
e Near Norwich.

SPECIES 3.

Subelliptical, round topped, not very pointed at the end; rays biporous on the top, pustulose on the base; pustulæ smaller than in the last species.
a Moushold.
b Taverham.
c Lexham.

SPECIES 4.

Oval, very pointed at one end, rounded at the other; rays biporous one third of the height downwards from the apex, very distinct on the base; ridge sharp; apex excentric towards the broad end.

Lexham.

The excentric situation of the apex, with the very pointed end, gives this species a peculiar appearance.

SPECIES 5.

Oval, pointed at one end, high; rays biporous two-thirds of the height downwards from the apex; areæ flattened.

a Wilts.

b Hickling.

Guildford.

SPECIES 6.

High, with flattish sides, top rounded or flattened; rays biporous downwards to the margin; height almost equal to the length.

a Holkham Park.

b Lexham.

SPECIES 7.

Three-fourths elliptical, the base, forming the section, rounded, contracted.

a Clayton Hill.

b Wilts.

SPECIES 8.

Broad-ovate, one end pointed; margin of the base undulated beneath with fourteen projections, the areolæ form five, the ridge one, and there are eight others in the areæ; a large depression on the upper part of the shell behind the apex.

Taverham.

SPINES OF ECHINI.

SPECIES 1.

Fig. 13, *Upper Chalk Plate.*

Long, slender, muricated.

Norwich.

SPECIES 2.

Clubshaped or gibbous, granulated in lines.

Kent.

Guildford.

SPECIES 3.

Clavicula cucumerina.
Long, of nearly equal thickness, granulated in quincuncial order.
Surry.

BONES.

TEETH.

SPECIES 1.

Fig. 14, *Upper Chalk Plate.*
Long, roundish, with two sharp ridges.
a Near Warminster.
b Norwich.

SPECIES 2.

One side flat, the other roundish ; two-edged.
a Wilts.
b Warminster.

SPECIES 3.

Fig. 15, *Upper Chalk Plate.*
Flattish, with serrated edges.
a Norwich.
b Near Warminster.

SPECIES 4.

Triangular, flat, keen-edged.
a Cherry Hinton.
b Near Warminster.

FISH PALATES.

SPECIES 1.

Fig. 11, *Upper Chalk Plate.*
Large, depressed; the middle elevated, deeply furrowed, the surrounding parts rough with small protuberances.
Near Warminster.

SPECIES 2.

The middle very high, deeply furrowed, the surrounding parts rough with small protuberances.
Near Warminster.

VERTEBRÆ OF FISH.

Fig. 16, *Upper Chalk Plate.*
Very small, concave at both ends.
North of Riegate.

GREEN SAND.

ZOOPHITA.

MADREPORA.

Fig. 15, *Green Sand, 2d Plate.*
Circular, depressed, radiated; radii very numerous on the upper surface, fewer on the under.
Chute Farm.

MILLEPORA.

Fig. 16, *Green Sand, 2d Plate.*
Branching, surface reteporous.
Chute Farm.

ALCYONIUM.

SPECIES 1.

Fig. 1, *Green Sand Plate.*
Funnel shaped, top concave.
Pewsey.
Warminster.
Devizes.

SPECIES 2.

Fig. 2, *Green Sand Plate.*
Doliform, small at each end.
Pewsey.

SPECIES 3.

Fig. 17, *Green Sand, 2d Plate.*
Round, with a central opening, and many minute pores on the surface.
Chute Farm.

ENCRINUS.

SPECIES 1.

Vertebræ pentagonal, with blunt projecting angles; alternately one large and three smaller, the middle one of the three larger than the other two.

Chute Farm.

SPECIES 2.

Vertebræ circular, alternately large and small, the large one twice as thick as the smaller. In this specimen, consisting of eight vertebræ, each terminating vertebra has five roundish oblique articulating surfaces protruded from it, probably the attachments of lateral branches.

Chute Farm.

TESTACEA.

UNIVALVIA.

TROCHUS.

SPECIES 1.

Volutions five or six, angular, the upper part flat, the side flattish or slightly depressed in the middle; base depressed, spirally furrowed; aperture wide, umbilicus small.

a Chute Farm.

b Near Warminster.

SPECIES 2.

Conical, sides straightish; volutions about six, slender, flattish beneath; striated; umbilicate; height equal to the width.

a Chute Farm.

b Rundaway Hill.

SOLARIUM.

Fig. 2, Green Sand, 2d Plate.

Much depressed; volutions angular, depressed, mouth wide.

Rundaway Hill.

TURRITELLA.

Fig. 5, Green Sand Plate.

Volutions about fifteen, covered by small crenulated lines; upper edge sharply crenated. About three larger crenulated lines with several lesser ones interposed.

Blackdown.

AMPULLARIA.

SPECIES 1.

Oblong, smooth; spire acute, with four or five volutions, sharp at their upper edge; aperture half the length of the shell. Length three-fourths of an inch.

Rundaway Hill.

TUBULAR IRREGULAR UNIVALVES.

VERMICULARIA.

SPECIES 1.

Vermicularia concava. *Min. Conch.*

Discoid, involute, concave on one side; the last volution but slightly attached: volutions about four.

Near Warminster.

SPECIES 2.

Vermicularia umbonata. *Min. Conch.*

Discoid, involute, umbonated above, concave beneath, the smallest whorl concealed in the umbo: whorls two or three.

Near Warminster.

SPECIES 3.

Discoid, involute, volutions few, surface rough with small hollows.

SERPULA.

SPECIES 1.

Cylindrical, straightish, with distant annular risings; opening round.

a Chute Farm.

b Near Warminster.

SPECIES 2.

Cylindrical, small, contorted.

Chute Farm.

MULTILOCULAR UNIVALVES.

BELEMNITES.

Elongated, with a slight furrow from the apex.

Chute Farm.

AMMONITES.

NO KEEL OR FURROW ON THE BACK.

SPECIES 1.

Ammonites Nutfieldiensis. *Min. Conch.?*

It seems more depressed.

Stourhead.

A KEEL ON THE BACK.

SPECIES 2.

Ammonites tuberculatus.

Involute, keeled; inner volutions half exposed, surface rough with large tubercular projections; tubercles oblong on the inner margin, furcating into two subnodose ridges; back wide, edged by large pointed knobs; keel entire.

a Rundaway Hill.

b Chute Farm.

A FURROW ON THE BACK.

SPECIES 3.

Extremely depressed; back narrow, concave, with crenulated edges; inner volutions concealed.

Longest diameter half an inch.

Chute Farm.

TURRILITES.

SPECIES 1.

Turrilites costata. *Min. Conch.*

Volutions of the spire beset with short ribs on the upper part, beneath which are two rows of small tubercles.

Chute Farm.

SPECIES 2.

Volutions of the spire convex, ornamented with two rows of close and small tubercles.

Evershot.

BIVALVIA.

INEQUIVALVED BIVALVES.

PECTUNCULUS.

Fig. 6, Green Sand Plate.

Orbicular, one side straightish, convex; beak rather prominent, acute; covered by many obscure furrows and minute transverse striæ; margin serrated within.

Teeth of the hinge about sixteen, rather large; the straightest side has a slight depression near the beaks, which makes it sublobate.

Blackdown.

CUCULLÆA.

SPECIES 1.

Fig. 10, *Green Sand Plate.*

Ovate, transverse, one side straightish; covered with numerous longitudinal decussated striæ; breadth less than three fourths of an inch.

Blackdown.

TRIGONIA.

SPECIES 1.

An inside cast of the trigonia of the Portland rock?

Chute Farm.

SPECIES 2.

An inside cast of a small species, or perhaps of an astarte; two muscular impressions.

Rundaway Hill.

SPECIES 3.

Trigonia clavellata. *Min. Conch.*

An impression of the outside.

Near Danby Beacon.

CARDIUM.

Cardium Hillanum. *Min. Conch.*

Nearly circular, a little oblique, covered with numerous concentric striæ; anterior part straightish at the edge; longitudinally furrowed.

Rather wider than long, a little gibbous.

The longitudinal furrows on the anterior side occupy about one fourth of the surface.

Blackdown.

A large transverse bivalve with unequal sides; one side produced and flattish, the other truncated and gibbous; transversely striated; beaks turned toward the produced side: length two inches, breadth three.

Blackdown.

VENUS.

SPECIES 1.

Venus plana. *Min. Conch.*

Rather depressed, subcordate, slightly angular towards the anterior side; surface smooth; edge entire: rather wider than long.

Blackdown.

SPECIES 2.

Venus angulata. *Min. Conch.*

Fig. 3, Green Sand Plate.

Obtusely cordate, wider than long, smooth; margin entire; an angular rising on the anterior side, which is slightly truncated; larger hinge teeth placed at a curved angle of about sixty degrees.

In the posterior side of the hinge is a roundish hollow, receiving a tooth on the opposite valve.

Blackdown.

INEQUIVALVED BIVALVES.

DIANCHORA.

Dianchora striata. *Min. Conch.*

Oblique, ovate, triangular, beak prominent, free valve obscurely ribbed; margin sharp. The length and breadth almost equal.

Chute Farm.

CHAMA.

Chama haliotidea. *Min. Conch.*

Fig. 7, Green Sand, 2d Plate.

Oval, uneven, a deep curving hollow within the deepest valve, extending from the beaks around one side; margin thin, broad, slightly fringed, crenate within; muscular impression large; upper valve flat.

a Dilton.
b Black-dog Hill, near Standerwick.
c Teffont.
d Evershot.
e Stourhead.
f Alfred's Tower.
g Blackdown.

OSTREA.

SPECIES 1.

Ostrea crista galli, Cockscomb oyster.

a Chute Farm.
b Stourhead.
c Blackdown.

SPECIES 2.

Fig. 10, *Green Sand, 2d Plate.*

Oblong, gibbous, smooth; front wide, a lobe on the left side forming a short wing, right side arched; lower valve gibbous, narrowed towards the back, with a curved projecting beak: margin thin, sharp.

Pit of the hinge arched, and continued up the under side of the beak; muscular impression nearest to the left side; right edge thickened within into a projecting ledge, left edge almost reflected.

a Stourhead.

b Dinton Park.

c Tinhead.

SPECIES 3.

Oval; lower valve equally convex, or scarcely at all lobate, the beak slightly curved, small; upper valve very concave.

Stourhead.

PECTEN.

SPECIES 1.

Pecten quadricostata. *Min. Conch.*

Fig. 8, *Green Sand, 2d Plate.*

Triangular, nearly even, front semicircular, margin notched; convex valve ribbed, larger ribs six, three smaller between each: posterior auricle large.

Near Warminster.

SPECIES 2.

Pecten quinquecostata. *Min. Conch.*

Subtriangular, rather oblique, front semicircular, toothed; convex valve ribbed, principal ribs six, four smaller ones between each; upper valve sulcated, flat-toothed; surface finely striated transversely.

Chute Farm.

SPECIES 3.

Pecten sexcostata.

Subtriangular, oblong, front semicircular, margin toothed; convex valve ribbed, principal ribs six, five lesser ones between each; surface deeply hollowed between the principal ribs.

Chute Farm.

SPECIES 4.

Fig. 3, *Green Sand, 2d Plate.*

Orbicular, depressed, longitudinally undulated; nearly twenty principal ridges covered with short spinous tubercles, and four or five lesser tuberculated ridges alternating with each of the principal ones.

Length and breadth nearly equal, two or three inches.

Chute Farm.

SPECIES 5.

Fig. 9, *Green Sand, 2d Plate.*

Oblong, very depressed, front semicircular; covered with many acute crenated threads, and intermediate decussated striæ.

Valves almost equally convex; the threads about one-third more numerous on the longer valve than on the shorter.

Chute Farm.

SPECIES 6.

Suborbicular, much depressed, about fifteen acute smooth ribs; ears reticulated, triangular.

Longleat Park.

SPECIES 7.

Orbicular, very depressed, transversely striated, smooth; ears large, smooth, triangular.

a Longleat Park.

b Warminster.

TEREBRATULA.

NOT PLICATED.

SPECIES 1.

Terebratula biplicata. *Min. Conch.*

Oblong, gibbous; beak prominent, sides rounded, front straightish, when full grown elevated with two distinct risings at the angles.

Lines of growth strong, sometimes irregular.

a Chute Farm.

b Near Warminster.

c Rundaway Hill.

PLICATED.

SPECIES 2.

Wider than long, depressed, beak very acute, front elevated with one wave rounding into the sides; plaits numerous, entire.

The lateral plaits are blended with those on the middle of the shell; width less than an inch; upper valve most convex.

a Chute Farm.

b Warminster.

SPECIES 3.

Upper valve gibbous; front straightish, much elevated with one large wave rounding into the sides; plaits numerous, entire; beak acute, very projecting.

The lateral plaits are blended with those on the middle of the shell: width less than half an inch; lower valve very depressed. This species differs from SPECIES 4 of the chalk in the greater convexity of the upper valve, and in the very slight incurvation of the beak.

a Chute Farm.

*b*Warminster.

SPECIES 4.

Terebratula obsoleta. *Min. Conch.*

The plaits are sharper than in Mr. Sowerby's figure, and the elevation in the front is more angular.

Near Warminster.

SPECIES 5.

Fig. 6, *Green Sand, 2d Plate.*

Subtriangular, wider than long, plicato-striated; front elevated with one wave rounding into the sides; plicæ numerous, rounded; beak acute, perforation large, triangular? lower valve advanced much beyond the upper.

a Chute Farm.

b Warminster.

SPECIES 6.

Terebratula pectinata. *Min. Conch.*

Fig. 4, *Green Sand, 2d Plate.*

Orbicular, gibbose, plicato-striated; a flattish space extending from the front to the beaks; beak of the lower valve prominent, slightly incurved; back of the upper valve straight, with an incurved beak.

The plicæ are small, rounded, and often furcate, on which account they are not much larger at the front than at the beaks. Length not more than one inch.

a Chute Farm.

b Warminster.

SPECIES 7.

Terebratula lyra. *Min. Conch.*

Fig. 5, *Green Sand, 2d Plate.*

Oblong, convex, with diverging furcated plaits, beak of the lower valve greatly elongated, that of the upper valve short, incurved.

The length of the upper valve is equal to twice its width. The beak of the lower valve is equal in length to the upper valve.

Chute Farm.

ECHINI.

ANOCYSTI.

CIDARIS.

SPECIES 1.

Circular, depressed, rays biporous; twenty rows of articulated eminences, which are nearly equal on the base, but unequal above, the two rows in each areola diminishing faster than those in the areæ; edges of the mouth turned inwards.

Chute Farm.

SPECIES 2.

Rather pentagonal, depressed, top convex; rays biporous, two rows of small eminences in each areola, forming the angles, two rows of miliæ in each area; mouth large, its edges turned inwards.

Near Warminster.

SPECIES 3.

Circular, depressed, with thirty rows of miliæ, two rows in each areola, and four rows in each area; rays biporous.

Chute Farm.

SPECIES 4.

Cidaris diadema.

Fig. 11, *Green Sand, 2d Plate.*

Depressed, upper side convex, base rounded; ten rows of alternate mammellæ, two rows in each area; areolæ narrow, subflexuous, widest on the margin, each side bordered by a row of small eminences; mouth large, with ten rather unequal sides, and a deep notch at every angle.

a Chute Farm.

b Near Warminster.

A curious anal appendage may be observed on most of the specimens obtained from this Stratum; a kind of frill surrounds the vent, and a tubular body projecting from it.

SPECIES 5.

Rather high; ten rows of mammellæ, two rows in each area, five mammellæ in a row; areolæ narrow, flexuous; apertures large.

Chute Farm.

SPECIES 6.

Subglobose, circular or pentagonal; rays biporous, depressed; areæ swelling; surface rough with small points.

VAR. with both the upper surface and base flattish.

Chute Farm.

PLEUROCYSTI.

SPATANGUS.

RAYS PLACED IN FURROWS.

SPECIES 1.

Cordate, margin rounded; dorsal furrow large, dorsal ridge high; height of the shell two thirds of the length.

a Charmouth, high variety.

b Melbury.

SPECIES 2.

High, base flattish; rays short, indented; dorsal furrow wide, not deep, ridge rising very high; vent placed as high as the apex; length one third of an inch, height and breadth one fourth.

Chute Farm.

RAYS NOT PLACED IN FURROWS.

SPECIES 3.

Fig. 14, *Green Sand, 2d Plate.*

Base flat, top convex, or rather flattened; dorsal groove short; rays ten biporous lines diverging in pairs.

a Chute Farm.

b Near Warminster.

ECHINITES lapis cancri. *Leske.*

Fig. 13, *Green Sand, 2d Plate.*

Obtusely ovate, gibbous, broader at one end than at the other; vent above the margin at the broad end, placed over a small depression or furrow; rays biporous; mouth five angled, small, nearest the narrow end; base swelling.

Chute Farm.

CATOCYSTI.

CONULUS.

Fig. 12, *Green Sand, 2d Plate.*

Convexo-conical, circular, base concave in the middle; rays ten, diverging in five pairs, the two rays near togetherseparated by a flat rising; mouth ten-angular, vent oblong.

a Chute Farm.
b Near Warminster.

BONES.

TEETH.

SPECIES 1.

One side flat, the other round, two-edged.
Chute Farm.

SPECIES 2.

Conoidal, flat, keen edged.
Chute Farm.

BRICK EARTH.

TESTACEA.

UNIVALVIA.

MULTILOCULAR UNIVALVES.

BELEMNITES.

Fig. 4 *and* 5, *Brick Earth Plate.*

Small, fusiform, swelling below the apex. Diameter less than half an inch, generally about a fourth. In some specimens there appear two very small grooves from the apex on opposite sides of the shell.

a North of Riegate.
b Near Godstone.
c Near Grimston.
d Steppingley Park.
e Prisley Farm.
Leighton Beaudesert.
Westoning.

AMMONITES.

A KEEL ROUND THE BACK.

SPECIES 1.

Much depressed, inner volutions not much concealed; radii unequal, dividing from small knobs on the inner part of the volution, prominent on the outer edge, obscure on the middle.

Mesterham, out of a deep well.

A FURROW ROUND THE BACK.

SPECIES 2.

Fig. 1, *Brick Earth Plate.*

Depressed, volutions increasing quickly, back wide, concave, with deeply furrowed edges; radii numerous, sharp, bifurcating from an oblong tubercle on the inner margin; edges of the back alternately indented; aperture oblong, sides almost parallel: inner volutions half exposed.

a Near Godstone.
b Steppingley Park.
c Prisley Farm.

HAMITES.

Fig. 2, Brick Earth Plate.

Limbs straightish, a little depressed; outer part of the limb tubercled on both sides; each tubercle dividing into two small ribs crossing the inner part, two annular ribs between the tubercles. Greatest thickness about half an inch.

Near Grimston.

BIVALVIA.

INEQUIVALVED BIVALVES.

PERNA.

Perna aviculoides? *Min. Conch.*
Godstone.

OSTREA.

A lower valve; gibbous, small.
Steppingley Field.

INOCERAMUS.

Fibrous shell.
a Prisley Farm.
b Near Grimston.

ECHINI.

PLEUROCYSTI.

SPATANGUS.

Fig. 3, Brick Earth Plate.

Cordate, depressed, dorsal furrow not deep; rays long, deeply hollowed; mouth depressed.

Near Devizes.

VERTEBRÆ OF SOME FISH.

Very small, concave at both ends, slender in the middle; ends rather oval.

North of Riegate.

PORTLAND ROCK.

WOOD.

Woburn.
Fonthill.
Swindon.

ZOOPHITA.

MADREPORA.

Stars angular, concave, numerous; radiating lamellæ twenty or more, diverging in pairs or in small bundles.

Tisbury.

TESTACEA.

UNIVALVIA.

TROCHUS.

Conical, sides undulated; volutions about six, the upper part flattish.

Swindon.

TURRITELLA.

Fig. 2, *Portland Rock Plate.*

Turreted; volutions about twelve, the upper edge projecting into a crenulated rim, lower part rounded, with a row of small protuberances almost concealed by the next volution; middle of the spire with three or four obscurely crenated striæ; base with crenated striæ: aperture rhomboidal, acute at the upper and lower edges.

a Portland.
b Swindon.

NATICA.

Fig. 1, *Portland Rock Plate.*

Subglobose, smooth, volutions few.

Swindon.

MULTILOCULAR UNIVALVES.

AMMONITES.

NO KEEL OR FURROW ON THE BACK.

Ammonites Nutfieldiensis. *Min. Conch.*

Involute, volutions four or more, nearly concealed; radii numerous, prominent, with shorter intermediate ones over the rounding back: aperture obcordate.

Fonthill.

Swindon.

BIVALVIA.

EQUIVALVED BIVALVES.

TRIGONIA.

SPECIES I.

Ovate, with many small roundish tubercles in quincuncial order, which terminate before a longitudinal ridge bounding the smooth anterior side; posterior side rounded, anterior side narrow, straightish at the edge:

The small tubercles are often united on the posterior side into connected oblique ridges.

a Fonthill.

b Swindon.

c Chicksgrove.

d Teffont.

Garsington Hill.

SPECIES 2.

Much elongated transversely.

In the cast the anterior slope is straight or concave, showing a large protuberance in the situation of the muscular impression. The hinge seems to have projected less into the shell in this than in the other species.

Swindon.

CARDITA.

Quadrangular, gibbous, beaks placed at one of the angles, and hooked; transversely furrowed; anterior side flattened and acuminated.

Crockerton.

ASTARTE.

Astarte cuneata. *Min. Conch.*

Fig. 3, *Portland Rock Plate.*

Subcordate, acuminated, gibbous, with small transverse costæ, lunette cordate, margin entire within.

The cast of the inside is triangular and acuminated at the anterior side.

Swindon.

VENUS.

SPECIES 1.

Ovate, wider than long, sides almost equal, beaks rather hooked.
Pottern.

SPECIES 2.

Fig. 5, *Portland Rock Plate.*
Subcordate, wider than long, gibbous; smooth.
Chicksgrove.
Swindon.

UNIO.

A species resembling Unio uniformis *Min. Conch.* but seems wider in proportion, with smaller beaks.
Swindon.

SMALL BIVALVES.

Teffont, in thin cherty stone.
Lady Down.

INEQUIVALVED BIVALVES.

OSTREA.

Irregular, flattish, shell thin.
Swindon.

PECTEN.

Fig. 6, *Portland Rock Plate.*
Circular, depressed, with transverse squamous lines of growth; valves almost equal; ears large, rectangular.
Length more than three inches.
Chicksgrove.
Swindon.

OAKTREE CLAY.

TESTACEA.

UNIVALVIA.

TROCHUS.

Fig. *Oaktree Clay Plate.*

Conical, sides undulated; volutions convex, the upper part flattish, bounded by a spiral crenulated rim; reticulato-striated; base rather convex, without an umbilicus.

Variety. Volutions subtuberculated on the upper part.

North Wilts Canal.

TURBO.

Fig. *Oaktree Clay Plate.*

Oblong, subturreted; volutions of the spire six, covered with rows of small tubercles, of which two rows in the middle of the spire are larger than the others: aperture about half the length of the shell.

Length about an inch.

North Wilts Canal.

MELANIA.

Melania Heddingtonensis. *Min. Conch.*

Volutions of the spire eight or more; surface of each volution concave near the middle, with an obtuse-angled rising near the upper part; above three times as long as the diameter; lines of growth deep.

North Wilts Canal.

TUBULAR IRREGULAR UNIVALVES.

SERPULA.

SPECIES 1.

Cylindrical, straightish, with distant annular risings, smooth; diameter not more than half an inch; thickness of the shell one sixth of the diameter.

North Wilts Canal.

Bagley Wood Pit.

SPECIES 2.

Elongated, rough, with five longitudinal undulations crossed by transverse wrinkles; diameter one third of an inch; thickness of the shell one fourth of the diameter; opening round.

a Brinkworth Common.
b Hinton Waldrish.
c Portland.
North Wilts Canal.
Bagley Wood Pit.

MULTILOCULAR UNIVALVES.

NAUTILUS.

Umbilicus open ?; septa broad, with one wave toward the aperture; back flattened; shell thick; siphuncle about a third of the length of the septum from the inner side.

Longest diameter from four to twelve inches, depth almost equal to the diameter.

North Wilts Canal.

AMMONITES.

NO KEEL OR FURROW ON THE BACK.

SPECIES 1.

Depressed, acuminated at the outer edge, inner volutions half exposed; radii numerous, unequal, prominent on the inner part, then furcate, or with one to three intermediate ones over the outer half of the volution, curving toward the aperture.

Longest diameter less than an inch.
a Brinkworth Common.
b Portland.

SPECIES 2.

Gibbous, inner volutions half concealed, radii unequal, very small and numerous, sharp across the back; aperture wider than long, back broad.

Longest diameter less than an inch.
a Brinkworth Common.
b Portland.
Dun's Well, Silton Farm.

SPECIES 3.

Fig. *Oaktree Clay Plate.*

Depressed, volutions numerous, the inner ones two thirds exposed; radii numerous, large, furcating, smoothed over the back; outer volutions plain; aperture oval, indented.

North Wilts Canal.
Well near Swindon, Wilts and Berks Canal.

SPECIES 4.

Volutions concealing each other ?, thickest near the center; nearly smooth.

Well near Swindon, Wilts and Berks Canal.

SPECIES 5.

Volutions armed with two rows of large tubercles, those on each side of the back most projecting; inner volutions little concealed; mouth squarish. Greatest diameter four inches.

Well near Swindon, Wilts and Berks Canal.

A KEEL ON THE BACK.

SPECIES 6.

Volutions much depressed, with a high thin keel serrated on the edge; radii many, unequal, prominent about the middle, and on the edge of the back.

The tubercles on the edge of the back are more in number than those on the middle of the volution.

Well near Swindon, Wilts and Berks Canal.

BELEMNITES.

Rather four sided, elongated, tapering quickly to the apex.

Well near Swindon, Wilts and Berks Canal.

North Wilts Canal.

BIVALVIA.

EQUIVALVED BIVALVES.

MODIOLA.

Transversely oblong, gibbous diagonally from the beaks; posterior side blunt, rounded in front; beaks small, hooked; back straightish, rounded into the anterior end; lines of growth sharp: shell very thin. Depressed in front of the gibbous part.

North Wilts Canal.

TRIGONIA.

SPECIES 1.

Trigonia costata. *Min. Conch. Park.*

North Wilts Canal.

SPECIES 2.

Trigonia clavellata. *Min. Conch. Park.*

Triangular, rather wider than long, with ten or more oblique rows of tubercles; anterior side straight, with three longitudinal knotted ridges.

North Wilts Canal.

SPECIES 3.

Trigonia curvirostra.

Oblong, triangular, narrow towards the beak; anterior side depressed, bounded by a very sudden termination of the other more convex part of the shell; posterior side rounded from the beaks into an oblique straightish front; covered by small transverse furrows which are diminished in size on the anterior side; beaks much hooked from the anterior side.

Length about an inch.

North Wilts Canal.

CARDITA.

Subtriangular, gibbose; beaks prominent; anterior side narrow, straightish, bounded by a smooth rising extending from the beaks to one angle; closely and finely striated transversely, on the anterior side a few obscure longitudinal striæ, length not more than half an inch.

Well near Swindon, Wilts and Berks Canal.

North Wilts Canal.

CARDIUM.

SPECIES 1.

Circular, gibbose; beaks prominent, contiguous; faintly striated transversely. Length about an inch.

North Wilts Canal.

SPECIES 2.

Wider than long, gibbose; beaks prominent, incurved, placed near one side; the opposite side produced; many longitudinal ridges crossed by transverse furrows.

North Wilts Canal.

Well near Swindon, Wilts and Berks Canal.

ASTARTE.

Astarte ovata.

Fig. *Oaktree Clay Plate.*

Transversely oblong, depressed, anterior side lengthened; transversely striated; lunette elliptical; margin crenulated within; shell thick; beak acute, solid: a small pit beneath the posterior slope of the hinge.

North Wilts Canal.

MACTRA.

Transversely oblong, posterior side straightish up to the beaks, anterior side produced, gaping; transversely furrowed.

Length more than an inch, breadth almost twice the length.

The posterior side very narrow, almost at a right angle to the front, with an obtuse angle in the middle.

Well near Swindon, Wilts and Berks Canal.

North Wilts Canal.

TELLINA.

Transverse, ovate, beaks small; closely striated transversely; a slight inflexion on the more lengthened side; shell thin.

Length an inch and a half.

Well near Swindon, Wilts and Berks Canal.

INEQUIVALVED BIVALVES.

CHAMA.

SPECIES 1.

Oblong, lower valve deep, the left side somewhat lobate, beaks subinvolute; upper valve flat; margin entire within.

North Wilts Canal.

Well near Swindon, Wilts and Berks Canal.

Bagley Wood Pit.

SPECIES 2.

Chama striata.

Oblong, elongated, curved, longitudinally striated; striæ irregular.

Bagley Wood Pit.

North Wilts Canal.

OSTREA.

SPECIES 1.

Ostrea deltoidea. *Min. Conch.*

Fig. Oaktree Clay Plate.

Equivalved, flat, thin, deltoidal, with a deep sinus on one side, and a produced straight beak.

Kennet and Avon Canal at Seend.

North Wilts Canal.

Well near Swindon, Wilts and Berks Canal.

Bagley Wood Pit.

Wilts and Berks Canal near Shrivenham.

Even Swindon.

Near Wotton Basset.

SPECIES 2.

Ovate, hooked, depressed; shell thin, with large transverse undulations; beak pointed.

Length an inch and a quarter, breadth three quarters.

Well near Swindon, Wilts and Berks Canal.

SPECIES 3.

Obliquely ovate, surface even, smooth; lower valve deep, beak not projecting; upper valve flat : shell thin.

North Wilts Canal.

SPECIES 4.

Ostrea crista galli. Cockscomb oyster.

North Wilts Canal.

Bagley Wood Pit.

SPECIES 5.

Depressed, uneven, with many roundish longitudinal ribs, crossed by imbricated lines of growth, forming distant rugous projections; inside of the shell plain.

North Wilts Canal.

AVICULA.

Avicula costata.

Brinkworth Common.

PECTEN.

SPECIES 1.

Nearly circular, depressed, closely striated longitudinally ; beak rectangular, ears produced, straight, rough.

North Wilts Canal.

SPECIES 2.

Circular, gibbose, many rough longitudinal ridges with smaller ridges interposed.

North Wilts Canal.

SPECIES 3.

Depressed, with roundish imbricated ribs.

Brinkworth Common.

TEREBRATULA.

PLICATED.

Fig. *Oaktree Clay Plate.*

Large, globose, the plaits on the middle of the shell blending with those on the sides; beak acute, much incurved.

North Wilts Canal.

Well near Swindon, Wilts and Berks Canal.

Bagley Wood Pit.

CORAL RAG AND PISOLITE.

ZOOPHITA.

MADREPORA.

SPECIES 1.

Fig. *Coral Rag and Pisolite Plate.*

Dichotomous; branches rough, closely striated longitudinally; the upper ends rather concave, radiated, lamellæ rough.

a Steeple Ashton.
b Longleat Park.
c Stratton.
d Ensham Bridge.
Wotton Basset.
Banner's Ash.
Well near Swindon, Wilts and Berks Canal.
Shippon.
Bagley Wood Pit.
Stanton near Highworth.

AGGREGATED.

SPECIES 2.

Upper side round or ovate, convex in the middle, covered with many hollow stars not divided by septa; radii numerous, unequal; lower side closely striated from the place of attachment.

Steeple Ashton.

SPECIES 3.

Fig. *Coral Rag and Pisolite Plate.*

Upper side covered with large pentagonal or hexagonal radiated cells with a central knob; radii numerous, unequal; lower side closely striated from the place of attachment.

The upper side frequently cordate or heart-shaped.
Stanton near Highworth.
South of Bayford.
Shippon.
Bagley Wood Pit.
Banner's Ash.

Well near Swindon, Wilts and Berks Canal.
Steeple Ashton.

FASCICULATED.

SPECIES 4.

Madrepora flexuosa?

Spreading, branches cylindrical, striated longitudinally, rough with transverse wrinkles, concave at the end; radii unequal in length.

Heddington Common.
Wotton Basset.

TESTACEA.

UNIVALVIA.

TROCHUS.

SPECIES 1.

Depressed; volutions few, angular, upper part flattish, with a row of sharp projecting tubercles.

Derry Hill.

SPECIES 2.

Trochus. *Oaktree Clay.*
Sandford Church Yard. Cast of the inside.
South of Bayford. Cast of the inside, more depressed.

MELANIA.

SPECIES 1.

Melania striata. *Min. Conch.*
Fig. Coral Rag and Pisolite Plate.

Volutions of the spire six or more, with about sixteen rounding furrows, more distant in the concealed parts: about $2\frac{1}{2}$ times as long as the greatest width.

a Calne.
b Steeple Ashton.
Silton Farm.
Banner's Ash.
Well near Swindon, Wilts and Berks Canal.
South of Bayford.

SPECIES 2.

Melania Heddingtonensis. *Min.Conch.*
a Heddington Common.

b Steeple Ashton.
Silton Farm.
Well near Swindon, Wilts and Berks Canal.
South of Bayford.

TURBO.

Turbo. *Oaktree Clay.*
Fig. *Coral Rag and Pisolite Plate.*

Oblong, subturreted; volutions of the spire six, covered with rows of small tubercles, of which two rows in the middle of the spire are larger than the others: aperture about half the length of the shell.

Length not more than an inch.

a Longleat Park.
b Derry Hill.
c Steeple Ashton.
Banner's Ash.
Wotton Basset.
Bagley Wood Pit.

TURRITELLA.

Very long, sides straight, volutions smooth, even.

Derry Hill.

AMPULLARIA.

Fig. *Coral Rag and Pisolite Plate.*

Oblong, spire acute, convex, smooth; volutions four or five, aperture oblong, narrow in the upper part.

a Longleat Park.
b Marcham.
Kennington.
Silton Farm.
South of Bayford.
Hinton Waldrish.

HELIX.

Conoidal, short; volutions four, convex, smooth, aperture half the length of the shell, umbilicus small.

Longleat Park.

TUBULAR IRREGULAR UNIVALVES.

SERPULA.

Elongated, with five obscure longitudinal undulations, crossed by transverse wrinkles: thickness of the shell a fourth of the diameter; opening round.

a Longleat Park.
b Derry Hill, almost spiral.
c Steeple Ashton, almost spiral.
Shippon, almost spiral
Kennington.

MULTILOCULAR UNIVALVES.

BELEMNITES.

Elongated, rather four-sided; diameter less than an inch.
Wotton Basset.
Shippon.

BIVALVIA.

EQUIVALVED BIVALVES.

MODIOLA.

SPECIES 1.

Coated muscle.
Long oval, convex, oblique, transversely striated.
a Wilts.
b Dry Sandford.
Well near Swindon, Wilts and Berks Canal.
Shippon.
Sunningwell.
Bagley Wood Pit.

SPECIES 2.

A cast of the inside; long oval, depressed, sides straightish, anterior end widest.
Hilmarton.
Banner's Ash.

TRIGONIA.

SPECIES 1.

Trigonia costata.
Silton Farm.
South of Bayford.

SPECIES 2.

Trigonia curvirostra. *Oaktree Clay.*
Longleat Park,

ASTARTE.

Astarte ovata. *Oaktree Clay.*
South of Bayford.
Well near Swindon, Wilts and Berks Canal. Cast of the inside.
Shippon.
Kennington. Cast of the inside.

Inside casts of a small bivalve; roundish, depressed.
South of Bayford.

MYA. *Linn.*

Cast of a depressed wide shell,
South of Bayford.

INEQUIVALVED BIVALVES.

PLAGIOSTOMA.

SPECIES 1.

Oblique, deltoidal, anterior side straight, posterior side circular, rounded into the front; ears short; anterior one placed in a wide furrow; many longitudinal sharp threads decussated near the edge by a few lines of growth.

Length two inches and a quarter.
Heddington Common.

SPECIES 2.

Plagiostoma gigantea? *Min. Conch.*
A small variety. Length not two inches.
Westbrook.
Calne.
Banner's Ash.
Well near Swindon, Wilts and Berks Canal.

CHAMA.

Chama, Species 1. *Oaktree Clay.*
Westbrook.
Banner's Ash.
Sunningwell.
Kennington.
Sandford Churchyard.
South of Bayford.
Silton Farm.

Specimens apparently of the same species.
a Longleat Park.
b Derry Hill.

OSTREA.

SPECIES 1.

Ostrea crista galli.
Fig. *Coral Rag and Pisolite Plate.*

Convex, oblong, curved; with many very sharp unequal plaits proceeding from the beaks, and dividing along the middle of the shell; lines of growth distinct; beak of the lower valve subinvolute, attached.

a Wilts.
Derry Hill.
b Shotover Hill.
c Westbrook.
d Longleat Park.
South of Bayford.
Wotton Basset.

SPECIES 2.

Circular; lower valve convex, depressed, slightly lobate on the left side; beak much curved; upper valve flat or concave; transversely laminated.

Length about five inches.
Steeple Ashton.

PECTEN.

SPECIES 1.

Pecten fibrosus. *Min. Conch.*
a Longleat Park.
b Heddington Common.
Kennington.

SPECIES 2.

Oblong, depressed; with twenty or more smooth ribs, and minute transverse striæ.

Highworth.

Sunningwell.

Westbrook.

SPECIES 3.

Circular, flattish, transversely striated; with twenty or more longitudinal ribs crossed by high transverse ridges.

Kennington.

Calne.

ECHINI.

ANOCYSTI.

CIDARIS.

SPECIES 1.

Depressed, upper side convex; apparently without any rows of eminences; areolæ prominent.

Near Abbotsbury.

SPECIES 2.

Pentangular, depressed, with thirty rows of small mammellæ which are almost equal on the sides; the row on each side of the areæ smaller than the others; rays obliquely triporous: mouth ten-sided, with a large notch at each angle.

Hilmarton.

Wotton Basset.

SPECIES 3.

Fig. *Coral Rag and Pisolite Plate.*

Subglobose, with ten rows of articulated mammellæ, each a little sunk and surrounded by a ring of small points, two rows in each area, separated by numerous small points; rays double, biporous, subflexuous, enclosing two rows of small points: many distinct points on each side of the area: apertures large.

Hilmarton.

Well near Swindon, Wilts and Berks Canal.

SPECIES 4.

Cidaris diadema. *Linn.*

Hilmarton.

Calne.

SPECIES 5.

Globose or conical, the top rounded, base flattish ; ten rows of very prominent articulated mammellæ, two rows in each area, separated by two rows of small distinct eminences ; each side of the area bordered by a row of small distinct eminences ; rays subflexuous, biporous ; areolæ narrow, edged by two prominent rows of points enlarged on the margin, widening from the vent downwards to the margin ; vent small.

The mammellæ are not separated from each other by any eminences, but the smooth base of one mammella touches the base of another.

CLYPEUS.

Fig. Coral Rag and Pisolite Plate.

Oblong, subquadrangular ; base flattish, concave in the middle ; mouth small, five angled ; upper side convex, with a large deep furrow on one side from the apex to the margin ; rays ten biporous lines in five pairs, depressed on the base : apertures opposite, excentric from the furrowed end. Shell unequally covered with small granulæ, most numerous on the base.

The furrow on the upper side is slight near the apex, then very deep, expanding toward the widest end, indenting the margin.

a Meggot's Mill, Coleshill.

b Longleat Park.

c Hinton Waldrish.

SPINES OF ECHINI.

Subcylindrical, elongated, smooth.
Hilmarton.

Clavicula cucumerina.
Longleat Park.
Derry Hill.

Very much lengthened, muricated.
Westbrook.

CLUNCH CLAY AND SHALE.

TESTACEA.

UNIVALVIA.

ROSTELLARIA.

Volutions of the spire five or six, a row of small studs on each margin; beak short.
Dudgrove Farm.

TUBULAR IRREGULAR UNIVALVES.

SERPULA.

Serpula, Species 2. *Oaktree Clay.*
This specimen is adherent and twisting.
Steeple Ashton.

MULTILOCULAR UNIVALVES.

BELEMNITES.

SPECIES 1.

Fig. *Clunch Clay Plate.*

Large, squarish, quickly tapering to the apex; diameter one inch at the large end, length four or five inches.

a Dudgrove Farm.
b North Wilts.

SPECIES 2.

Very long, slender, tapering gradually to a sharp point, furrow large in the upper part; length five inches, diameter half an inch.

Dudgrove Farm.

AMMONITES.

NO KEEL OR FURROW ON THE BACK.

SPECIES 1.

Ammonites communis. *Min. Conch.*
This shell varies extremely, in some specimens the aperture is oval, in others wide.
In this stratum generally wide.
Whitby.

SPECIES 2.

Ammonites, Species 3. *Kelloways Stone.*

Holt.

A KEEL ROUND THE BACK.

SPECIES 3.

Fig. *Clunch Clay Plate.*

Inner volutions two thirds concealed; radii many, sharp, twice curved, tubercular twice on the inner margin, and again on the outer. Shell very much compressed.

a Thames and Severn Canal.

b Tytherton Lucas.

SPECIES 4.

Ammonites Walcotii. *Min. Conch.*

Depressed, volutions four, three fourths exposed, with a concentrate furrow; lunate undulations over half the sides; back with a keel between two furrows: aperture oblong, one third of the diameter.

Each volution divided into two parts by an obtuse furrow; inner half nearly smooth.

Whitby.

BIVALVIA.

EQUIVALVED BIVALVES.

TELLINA.

Wider than long, gibbous, smooth, one side produced into a short thick beak, the other rounded and blunt; beaks incurved.

Whitby.

INEQUIVALVED BIVALVES.

OSTREA.

Gryphæa dilatata. *Min. Conch.*

Circular, obscurely lobed; upper valve flat, lower valve hemispherical.

Fig. *Clunch Clay Plate.*

VARIETY 1. Not very convex, large and broad, beak not much curved over the hinge; obscurely lobed: length from three to five inches.

a Meggot's Mill, Coleshill.

b Derry Hill.

c Between Weymouth and Osmington.

d Tytherton Lucas.

Dudgrove Farm ; smaller, attached at the beak.

Variety 2. Lobate oyster; left side very much lobed, lower valve deep.
a Steeple Ashton.
b Bubdown.

AVICULA.

Avicula costata.
Dudgrove Farm.

TEREBRATULA.

Terebratula obsoleta? *Min. Conch.*
Dudgrove Farm.

KELLOWAYS STONE.

TESTACEA.

UNIVALVIA.

ROSTELLARIA.

Fig. *Kelloways Stone Plate.*

Volutions seven, or more, with two smooth angular carinæ, the upper one very prominent, the other smaller and concealed; many fine striæ.

Kelloways.

Wilts and Berks Canal.

TURRITELLA.

Turreted, volutions five or six, angular, with many crenulated lines; aperture roundish.

a Thames and Severn Canal.

b Dauntsey House, in stone.

c Kelloways.

MULTILOCULAR UNIVALVES.

NAUTILUS.

Subglobose, volutions rounded, diminishing fast, smooth; umbilicate; many fine striæ across the back; shell thin.

Kelloways.

BELEMNITES.

Very long, slender, tapering gradually to a sharp point, furrow large.

Wilts and Berks Canal.

AMMONITES.

NO KEEL OR FURROW ON THE BACK.

SPECIES 1.

Ammonites sublævis. *Min. Conch.*

Fig. *Kelloways Stone Plate.*

Globular, (rather depressed when young), inner volutions exposed within the umbilicus, which is deep, undulated, and has an angular edge: septa numerous, with five principal undulations.

Aperture very wide, semicircular, truncated at the sides.

a Kelloways.

b Ladydown Farm.

Christian Malford.

SPECIES 2.

Fig. *Kelloways Stone Plate.*

Volutions six or more, half concealed, back round; radii numerous, unequal, united on the inner part of the volution by pairs or more into compressed oblong tubercles, sometimes with shorter intermediate radii over the back; aperture oblong, widest in the inner part, inner angles rounded.

Diameter about two inches.

a Kelloways.

b Dauntsey House.

c Wilts and Berks Canal.

d Kennet and Avon Canal.

SPECIES 3.

Volutions two thirds concealed, gibbous, with oblong sharp ridges on the inner part, from which proceed very numerous small radii; back wide, rounded; aperture obcordate, more than one third of the diameter in length, and as wide as long.

Diameter four inches.

Christian Malford.

SPECIES 4.

Ammonites Calloviensis. *Min. Conch.*

Fig. *Kelloways Stone Plate.*

Involute, subumbilicate, rather depressed; volutions about five, three fourths concealed, back flat; radii small, very numerous, alternately one long one prominent near the inner edge and from two to five short, obscure in the latter whorls of old shells; aperture roundish when young, deltoid with the angles truncated when old.

a Kelloways.

b Wilts and Berks Canal.

VARIETY. Depressed, volutions more than half concealed, radii twice prominent on the inner part, with from two to five shorter ones over the back.

Tytherton Lucas.

SPECIES 5.

Thickish, subumbilicate; radii twice very prominent or tuberculated on the inner part of the volution, with from two to five shorter ones on the outer part; back flat, edged by many small tubercles.

Tytherton Lucas.

SPECIES 6.

Gibbose, subumbilicate; volutions nearly concealed; radii close, alternately divided, tuberculated on the inner and outer margins; back flat.

Diameter one third of an inch.

Kelloways.

SPECIES 7.

Volutions four or five, exposed, with pointed tubercles on the inner part dividing into two or three sharp radii; a slight depression along the back. Tubercles more obscure in the outer volutions.

Tytherton Lucas.

BIVALVIA.

EQUIVALVED BIVALVES.

ARCA.

Much wider than long, ovate, depressed, with small transverse furrows; sides unequal; beaks small; shell thin.

Thames and Severn Canal.

CARDITA?

SPECIES 1.

Transverse, ovate, convex, subequilateral, with small transverse furrows, margin thick; breadth above an inch.

Kelloways.

SPECIES 2.

Subtriangular, gibbous, beaks prominent, anterior side narrow, straightish, bounded by a smooth rising extending from the beaks to one angle; closely and finely striated, both longitudinally and transversely, margin crenulated within.

Length about half an inch.

a Thames and Severn Canal.

b Dauntsey House.

c Kelloways.

MYA. *Linn.* UNIO? *Lam.*

Twice as wide as long, with obscure transverse undulations; anterior side produced, acuminated, posterior side blunt, front margin not much arched. Beaks rather prominent.

Wilts and Berks Canal.

Thames and Severn Canal.

INEQUIVALVED BIVALVES.

OSTREA.

Gryphæa dilatata. *Min. Conch.*

Fig. *Kelloways Stone Plate.*

VARIETY. Lobate oyster. Left side very much lobed, lower valve deep.

a Kelloways.

b Wilts and Berks Canal.

c Ladydown on the Biss River.

d Bruham Pit. Experiment for Coal.

AVICULA.

Avicula costata.

Kelloways.

TEREBRATULA.

Terebratula ornithocephala. *Min. Conch.*

Fig. *Kelloways Stone Plate.*

Ovato-rhomboidal; depressed when young, elongated and gibbous when old; front straigh bounded by two obtuse lateral depressions, similar in each valve.

a Thames and Severn Canal.

b Dauntsey House.

c Kelloways.

d Wilts and Berks Canal.

CORNBRASH.

ZOOPHITA.

PENTACRINUS.

Vertebræ thin, with very projecting angles.
a Wick Farm.
b Pipehouse.
c Well at Seagry.

TESTACEA.

UNIVALVIA.

VOLUTA.

Subcylindrical, smooth, spire short; aperture widest at the base.
Near Norton.
North side of Wincanton.

TURBO.

Volutions few, angular in the lower part.
Sheldon.

TURRITELLA.

Sides straight, volutions smooth, even, with a small ridge below the upper edge.
a Melbury.
b Lullington.

ROSTELLARIA.

Rostellaria. *Kelloways Stone.*
Melbury.

AMPULLARIA.

Depressed, small, volutions few.
Lullington.

NATICA?

Fig. 1, *Cornbrash Plate.*

Volutions three or four, the outer one very large, spire short, with an obtuse furrow beneath the projecting upper edge of the volutions.

Aperture semicircular, almost four fifths as long as the whole shell.

a Road.

b Sleaford.

Wick Farm.

TUBULAR IRREGULAR UNIVALVES.

SERPULA.

SPECIES 1.

Cylindrical, straightish, enringed with small projections, diameter from one fourth to one third of an inch.

a Melbury.

b Closworth.

c Farley.

d Sheldon.

SPECIES 2.

Cylindrical, small, contorted, lines of growth distinct.

a Trowle.

b Holt.

c Sheldon.

MULTILOCULAR UNIVALVES.

AMMONITES.

NO KEEL OR FURROW ON THE BACK.

Ammonites discus. *Min. Conch.*

Fig. 2, *Cornbrash Plate.*

Discoid, smooth, outer edge acuminated; aperture sagittate, half the diameter of the shell in length, and one sixth in breadth: volutions concealing each other, septa numerous, largely undulated.

a Closworth.

b Road.

Southwest of Wincanton.

BIVALVIA.

EQUIVALVED BIVALVES.

MODIOLA.

Fig. 3, *Cornbrash Plate.*

Transversely oblong, most convex diagonally from the beaks; posterior side rather blunt, rounded in front; back straightish, rounded into the anterior end; beaks small; lines of growth sharp; shell very thin.

Depressed in front of the gibbous part.

a Closworth.

b Wick Farm.

c Holt.

UNIO.

SPECIES 1.

Depressed, triangular, broader than long, anterior and posterior sides acuminated, front margin straightish, length from half to three fourths of an inch.

a Norton.

b Near Tellisford.

SPECIES 2.

Twice as wide as long, much depressed, beak a little nearest one side, the other produced, front margin straight.

Length three fourths of an inch.

This much resembles Unio acutus. *Min. Conch.*

Melbury.

SPECIES 3.

Fig. 7, *Cornbrash Plate.*

Twice as wide as long, depressed, beaks nearest to one end, the other produced, front not much arched; transversely furrowed.

a North Cheriton.

b Road.

c Draycot.

d Maisey Hampton.

e Sleaford.

f Southwest of Tellisford.

g Sattyford.

SPECIES 4.

Twice as wide as long, ovate, anterior side subacuminate, posterior side rather blunt; beaks notprominent; anterior slope arched, front straightish; transversely furrowed.

Down Ampney.

TRIGONIA.

SPECIES 1.

Trigonia clavellata. *Min. Conch. Park.*

a Melbury.

b Woodford.

SPECIES 2.

Trigonia costata. *Min. Conch. Park.*

Fig. 4, *Cornbrash Plate.*

VARIETY with very sharp ribs.

North side of Wincanton.

Wick Farm.

CARDIUM.

Fig. 6, *Cornbrash Plate.*

Subglobose, ovate, front and posterior sides rounded, anterior slope straight; surface undulated with about ten large ridges crossed by transverse furrows; beaks at one end, very protuberant, incurved.

Length from two to four inches.

a Road.

b Elmcross.

c Wick Farm. A smaller VARIETY; the posterior side bounded by a very prominent ridge forming an angle in the contour: front straightish.

d Sleaford.

e Woodford.

f Near Peterborough.

MYA?

Twice as wide as long, beaks at one end which is truncated and gibbous, the other produced, acuminated; front not much arched, anterior slope straight; transversely furrowed.

a Redlynch.

b Near Tellisford.

VENUS?

SPECIES 1.

Fig. 5, *Cornbrash Plate.*

Subcordate, gibbose, wider than long, anterior side subacuminated, posterior side blunt; beaks prominent, hooked, subincurved; transversely striated.

a Trowle.
b Sheldon.
Southwest of Wincanton.

Var. Subcircular or deltoidal, gibbose, posterior side straightish, extended to a point, beaks very much hooked above the posterior side.

a Trowle.
c Norton.

SPECIES 2.

Circular, depressed, with transverse furrows.
North of Latton.

SPECIES 3.

Small, depressed, beaks not curved, one side produced.
Sheldon.

SPECIES 4.

Small, circular, convex, subequilateral, with transverse sharp costæ.
Wick Farm.

INEQUIVALVED BIVALVES.

LIMA.

SPECIES 1.

Trigonal or semicircular, oblique, with short ears; one side straightish and lengthened, the other rounded; covered with many longitudinal sharp plaits.

a Melbury.
b Closworth.
c Norton.
d Wick Farm.

SPECIES 2.

Lima gibbosa. *Min. Conch.*
North side of Wincanton.

OSTREA.

SPECIES 1.

Ostrea Marshii. *Min. Conch.*

Oblique, both valves deeply plaited in seven or eight angular diverging undulations; edge thick, flatted.

a Sleaford.
b Woodford.

SPECIES 2.

Oblong, lower valve convex, upper valve concave; surface irregularly undulated or imbricated.

a Melbury.
b Closworth.
c Sheldon.
d Draycot.
e Woodford.
f Norton.

SPECIES 3.

Oblong, upper valve concave, lower valve convex; with longitudinal elevated unequal threads.

Woodford.

AVICULA.

SPECIES 1.

Avicula echinata.

Fig. 8, *Cornbrash Plate.*

Subcircular, convex, beak rather prominent, with a short wing or ear from one side; surface covered with near thirty small threads roughened by projecting lines of growth.

The length seldom exceeds two thirds of an inch.

a Closworth.
b North Cheriton.
c Lullington.
d Trowle.
e Sheldon.
f Draycot.
g Norton.
h Stony Stratford.
i Southwest of Tellisford.
North side of Wincanton.
Southwest of Wincanton.

The under valve only is here described, for the upper valve has not been observed among any of these specimens.

SPECIES 2.

Avicula costata.
Stony Stratford.

PECTEN.

Pecten fibrosus? *Min. Conch.*

Depressed, circular, with a rectangular beak; valves unlike, one valve with about ten broadish diverging grooves and numerous sharp concentric striæ, the other valve with about ten diverging ribs and many larger concentric scales or tiles; ears equal, margin undulated internally.

The ribs in one valve are alternately larger and smaller answering to the grooves in the other.

Length about an inch.

a Melbury.

b Sheldon.

c Woodford.

North side of Wincanton.

Southwest of Wincanton.

TEREBRATULA.

NOT PLICATED.

SPECIES I.

Terebratula intermedia. *Min. Conch.*

Obscurely five-sided, rather depressed, smooth, larger valve most convex, front margin undulated; three depressions in the smaller valve and two in the larger.

a Melbury.

b Bruham.

c Lullington.

d Road.

e Trowle.

f Maisey Hampton.

g Holt. A large specimen, two inches in length.

SPECIES 2.

Terebratula digona. *Min. Conch.*

Fig. 9, *Cornbrash Plate.*

Triangular, oblong, gibbous; beak prominent; sides rounded; front either convex or concave, when old bounded by two prominent angles alike in each valve.

This species is very variable in form, but specimens from this stratum are generally of the gibbous and shorter variety with the front margin slightly convex between the angles.

a Closworth.

b Redlynch.

c Trowle.

d Wick Farm.
e Sheldon.
f Latton.
g Woodford.

PLICATED

Terebratula obsoleta? *Min. Conch.*

Nearly round, gibbous, plaited; middle of the front a little elevated by seven plaits; sides having from seven to eleven plaits; beak projecting.

a Closworth.
b Draycot.
c Wick Farm.
Southwest of Wincanton.
North side of Wincanton.

ECHINI.

ANOCYSTI.

CIDARIS.

Cidaris, Species 2. *Clay over the upper oolite.*

a Melbury, with smaller mammellæ.
b Sheldon.
c Norton.
d Southwest of Tellisford. VARIETY with smaller mammellæ.
e Wick Farm.

CLYPEUS.

Clypeus. *Coral Rag and Pisolite.*

a Bruham.
b Wick Farm.
c Wraxhall.
d Sleaford.
e Trowle.
f Southwest of Tellisford.
Southwest of Wincanton.

CATOCYSTI.

CONULUS.

Conulus. *Fuller's Earth Rock.*

Nearly circular, convexo-conical, base concave; covered by small prominences particularly on the base of the shell; rays ten biporous lines diverging in five pairs, pores very close on the upper part, more distant on the base.

Woolverton.

Southwest of Wincanton.

FOREST MARBLE.

ZOOPHITA.

MADREPORA.

AGGREGATED.

SPECIES 1.

Circular, upper part convex, with large unequal radiated cells; lower part flat, closely striated from the center.

A transverse semicircular ostrea adhering to the lower part.

Laverton.

SPECIES 2.

Upper side flat, with large close cells; lower side convex, closely striated.

Stunsfield.

TESTACEA.

UNIVALVIA.

PATELLA.

Patella rugosa. *Min. Conch.*

Fig. 1, *Forest Marble Plate.*

Depressed, obovate, radiated; apex eccentric, depressed, slightly recurved; back concave above, with reflected undulations.

a Minching Hampton Common.

b Hinton.

ANCILLA.

SPECIES 1.

Fig. 2, *Forest Marble Plate.*

Subcylindrical, smooth, volutionssquare and subcrenulated at the upper edge; spire short, turreted; aperture two thirds of the length, widest at the base.

Farley.

SPECIES 2.

Fusiform, spire short, acute, base pointed, volutions smooth, slightly angular at the upper edge; aperture acute at the ends, widest in the middle, rather more than two thirds of the length.

Wincanton.

TURRITELLA.

SPECIES 1.

Slender, volutions many, angular, concave beneath the upper edge.

Marston, near Frome.

ROSTELLARIA.

Fig. 3, Forest Marble Plate.

Elongated, spire with longitudinal ridges acute on the upper edge, and transverse striæ. Length not more than three fourths of an inch.

Poulton.

TUBULAR IRREGULAR UNIVALVES.

SERPULA.

Cylindrical, small, contorted.

Farley.

BIVALVIA.

EQUIVALVED BIVALVES.

TRIGONIA.

SPECIES 1.

Trigonia costata. *Park. Min. Conch.*

The ribs very sharp, length one inch.

Wincanton.

SPECIES 2.

Triangular, wider than long, depressed, anterior side straight, posterior side rounded; with ten or more rather oblique rows of small tubercles, twenty or more tubercles in a row; two or three tuberculated ridges on the posterior side.

Length one inch.

Stunsfield.

VENUS. *Linn.*

Cast of the inside:
Transversely ovate, subequilateral; length one inch.
Road Lane.

MYA.

Transverse, gibbous, breadth three times the length; anterior end elongated, pointed, gaping for half the breadth; beaks small, one sixth of the breadth from the posterior end; transversely striated; thickness equal to the length, shell thin.
Stunsfield.

INEQUIVALVED BIVALVES.

OSTREA.

SPECIES 1.

Ostrea crista galli.
Stunsfield.

SPECIES 2.

Lower valve convex, oblique, widest and deepest near the front, flattened and attached toward the back.
Orchardleigh.

SPECIES 3.

Fig. 4, Forest Marble Plate.
Oval, flattish, depressed; upper valve flat, surface longitudinally wrinkled.
Pit oblong; shell hollow beneath the pit.
The pit of the upper valve is formed by the projection of the beak. Length one inch.
a Wincanton.
b Road, Coal Experiment.

PECTEN.

SPECIES 1.

Fig. 5, Forest Marble Plate.
Large, convex, eared, ribbed longitudinally, with forty or fifty wide smooth ribs; striated transversely, striæ obscure on the ribs.
Length, two inches and a quarter.
The ribs are obscure toward the beaks.
a Siddington.
b Foss Cross.

SPECIES 2.

Large, oblong, depressed, sides rather unequal, ribbed longitudinally with forty or fifty large ribs, often with an intermediate smaller one.

Length, three inches and three quarters.

Laverton.

SPECIES 3.

Pecten fibrosus. ? *Min. Conch.*

The diverging furrows are only near the front, and are striated, the intermediate spaces plain.

Stunsfield.

SPECIES 4.

Fig. 6, *Forest Marble Plate..*

Circular, depressed, striated longitudinally; striæ numerous, fine, dividing; intersecting near the beaks? ears large, rough.

Farley.

SPECIES 5.

Circular, depressed; transversely imbricated, faintly striated longitudinally; ears equal, rough; length two inches.

Farley.

SPECIMENS OF THE STONE

Contain

ROSTELLARIA.

HELIX.

Smooth, conoidal, obtuse.

Farley.

VENUS.

An oval transverse striated shell.

Farley.

OSTREA.

Rough, irregular.

Farley.

Wincanton.

SPINES OF ECHINI.

Slender, subcylindrical, smooth.

Farley.

BONES.

A long slender bone. Length, four inches and a half.
Stunsfield.

TEETH.

BUFONITÆ.

SPECIES 1.

Fig. 7, *Forest Marble Plate.*
Oval.
a Stunsfield.
b Pickwick.

SPECIES 2.

Fig. 8, *Forest Marble Plate.*
Circular: diameter a quarter of an inch.
a Stunsfield.
b Pickwick.
c Didmarton.

OBLONG, POINTED.

SPECIES 3.

Fig. 11, *Forest Marble Plate.*
Curved, tapering, round, with two sharp edges. Diameter from three sixteenths to one third of an inch, length five times the diameter.
a Stunsfield.
b Pickwick.

SPECIES 4.

Curved or oblique, flattened. Length one inch, diameter three sixteenths of an inch.
Stunsfield.

SPECIES 5.

Curved, a little flattened near the point, cylindrical at the base. Length three fourths of an inch, diameter one twelfth.
Stunsfield.

SPECIES 6.

Flattish, triangular, with two sharp edges and two small lateral teeth, one on each side. Length, half an inch.

a Stunsfield.
b Pickwick.

PALATES OF FISH.

SPECIES 1.

Fig. 9, *Forest Marble Plate.*

Foursided, angles acute and obtuse, depressed; sides parallel, one of them defined by an edge; surface uneven, covered with very small hollows.

Pickwick.

SPECIES 2.

Elongated, with parallel flattened sides, one of them defined by an edge; surface uneven, covered with small hollows.

Pickwick.

SPECIES 3.

Fig. 10, *Forest Marble Plate.*

Deltoidal or cap formed, much elevated in the middle; smooth.

Pickwick.

CLAY over the UPPER OOLITE.

ZOOPHITA.

TUBIPORA.

Roundish, lengthened, composed of very small angular closely aggregated tubes.
Broadfield Farm.
Farley.

MADREPORA.

SPECIES 1.

Madrepora porpites. *Linn.*
Broadfield Farm.

SPECIES 2.

Circular, convex and uneven above, with many equal concave stars.
Hinton.

MILLEPORA.

Fig. 5, Clay over the Upper Oolite Plate.

Ramose, branches round, composed internally of layers or coats of tubes opening into reticulating angular cells on the surface; branches undulated, sometimes spirally.
Broadfield Farm.
Farley.
Hinton.
Pickwick.
Westwood.

ENCRINUS.

Pear Encrinus.
Fig. 1, 2, and 3, Clay over the Upper Oolite Plate.

Bradford.	Heads, and vertebræ.
Farley.	vertebræ.
Hinton.	vertebræ.
Winsley.	vertebræ.
Pickwick.	vertebræ.

PENTACRINUS.

Vertebræ, with blunt projecting angles.
Towcester.

TESTACEA.

UNIVALVIA.

TROCHUS.

SPECIES 1.

Conical, sides undulated; volutions four or five, angular, the upper part flat; the lower edge keeled; striated spirally, on the sides a few oblique threads across the spire.
Bradford Lock.

SPECIES 2.

Short, conical, striated; volutions four or five, lower edge thin, strongly crenated; lines of growth distinct; umbilicate.
Bradford Lock.
Broadfield Farm.

TURRITELLA.

Long, slender, a spiral groove in the lower part of the volutions.
Farley.

TUBULAR IRREGULAR UNIVALVES.

SERPULA.

SPECIES 1.

Elongated, conical, twisting at the small end, quadrate with four projecting crenated angles or ridges; opening round.
a Westwood.
b Farley.
c Broadfield Farm.

SPECIES 2.

Large, cylindrical, straightish,s mooth.
Farley.

MULTILOCULAR UNIVALVES.

BELEMNITES.

Small, slender.
Stoford.

BIVALVIA.

EQUIVALVED BIVALVES.

MODIOLA.

SPECIES 1.

Mytilus tunicatus, or coated muscle. Fistulana. *Lam.*
Combhay.

SPECIES 2.

Small, elongated, widest in front; one side straightish and gibbous, the other rounded; longitudinally striated, striæ unequal in length.
Farley.

SPECIES 3.

Much elongated, curved, ends acuminated; most convex near the beaks; on the outer side of the arch near the beak a fold or constriction; lines of growth sharp.
Farley.

Modiola, *Cornbrash.*
Westwood.

TRIGONIA.

Trigonia costata. *Min. Conch.*
Var. The anterior side wider than the posterior; length two inches and a quarter, breadth one inch and a half.
Hinton.

VENUS?

SPECIES 1.

Inside casts of a small species, flattish, sides almost equal.
Combhay.

SPECIES 2.

Ovate, wider than long, gibbous; depth equal to the length; beaks large, hooked; transversely striated.

Poulton Quarry, Bradford.

INEQUIVALVED BIVALVES.

PLAGIOSTOMA.

SPECIES 1.

Fig. 7, *Clay over the Upper Oolite Plate.*

Ovate, depressed, front semicircular, back straight; closely striated longitudinally and finely decussated; ears small, one of them placed in a wide deep furrow; beaks flattened.

The longitudinal striæ are rather unequally distant on the margin and are finely decussated.

Bradford.

SPECIES 2.

Oblong, one side straight, the other semicircular; deeply striated longitudinally, striæ forty or more, decussated; ears small.

There appears to be a projecting flat surface beneath the beaks.

Farley.

VARIETY.

With more numerous striæ.

Combhay.

CHAMA.

SPECIES 1.

Chama crassa.

Fig. 6, *Clay over the Upper Oolite Plate.*

Very long, narrow, deep around one side, smooth; beak subinvolute; upper valve flat, tooth of the hinge small, blunt; margin entire within.

Length almost three times as great as the breadth.

Stoford.

SPECIES 2.

Oblong, gibbous, lobate in front; beak subinvolute; upper valve flat; margin entire within.

Length one inch and three eighths, breadth four fifths of an inch.

a Stoford.

b Combhay.

SPECIES 3.

Gibbous, beaks greatly curved, pit of the hinge large and deep; margin entire within.
Breadth four fifths of the length.
Pickwick.

OSTREA.

SPECIES 1.

Ostrea crista galli. Cock's comb oyster.
a Farley.
b Combhay.
c Hinton.

SPECIES 2.

Circular, depressed, longitudinally plaited; plaits unequal, often dividing; lines of growth distinct; beak central.
Farley.

SPECIES 3.

Ostrea acuminata. *Min. Conch.*
Depressed, very long, curved, with large subimbricated transverse waves beneath; beaks and front acuminated; upper valve nearly even.
Two or three times as long as wide.
Bradford.

PECTEN.

SPECIES 1.

Pecten fibrosus? *Min. Conch. Cornbrash.*
Farley.

SPECIES 2.

Circular, almost flat, smooth, ears small.
Farley.

AVICULA.

Avicula costata.
Fig. 8, *Clay over the upper oolite Plate.*
Oblique, subdepressed, left side extended, acuminated; wings unequal, the right wing short; valves unequal, lower valve convex, with about eight smooth ribs, and undulated internally, upper valve flat, with diverging striæ.

a Bradford,
b Hinton.
c Winsley.

TEREBRATULA.

NOT PLICATED.

SPECIES 1.

Terebratula digona. *Min. Conch.*

Fig. 9, *Clay over the upper oolite Plate.*

Triangular, oblong, gibbous; beak prominent; sides rounded; front either convex or concave, when old bounded by two prominent angles alike in each valve.

This species is very variable in form, but specimens from this stratum are generally of the lengthened variety with straightish flattened sides.

a Farley.
b Stoford.
c Bradford.
d Winsley.
e Pickwick.

SPECIES 2.

Obscurely five sided, rather depressed, smooth; larger valve most convex; front margin undulated, with three depressions in the smaller valve and four in the larger.

Length more than an inch.

a Farley.
b Stoford.
c Combhay.
Winsley.

PLICATED.

SPECIES 3.

Terebratula obsoleta? *Min. Conch.*

Nearly round, gibbous, plaited; middle of the front a little elevated with from six to eight plaits; sides having from nine to twelve plaits; beak projecting.

Depth about two thirds of the length.

Length from half an inch to an inch and a quarter.

a Farley.
b Westwood.
c Winsley.
d Pickwick.

SPECIES 4.

Oval, convex; with many diverging sharp plaits furcating from near the beaks; front margin not elevated, beak prominent.

Farley.

SPECIES 5.

Terebratula reticulata.

Fig. 10, *Clay over the upper oolite Plate.*

Obtusely five sided, gibbous; with minutely echinated or reticulated striæ; surface uneven; larger valve with two very prominent folds continued from the front to the beak, and three large depressions, forming large angular notches on the front; upper valve with corresponding undulations.

a Farley.

Bradford.

Stoford.

b Hinton.

c Winsley.

d Pickwick.

ECHINI.

ANOCYSTI.

CIDARIS.

SPECIES 1.

Circular, depressed, top convex; rays biporous; two rows of small eminences in each areola, two rows of miliæ in each area; mouth large, its edges turned inwards.

Broadfield Farm.

SPECIES 2.

Circular, rather depressed, with ten rows of contiguous mammellæ, (near the vent the first mammellæ in the rows are encircled), two rows in each area, separated by curvilinear rows of small eminences partly surrounding the mammellæ; areolæ with minute points encircling two rows of eminences; mouth large, decagonal, with a deep notch at every angle.

a Pickwick.

b Farley.

CLYPEUS.

Clypeus. *Coral Rag and Pisolite.*

Broadfield Farm.

UPPER OOLITE.

ZOOPHITA.

MADREPORA.

SIMPLE OR COMPOSED OF ONE SINGLE STAR.

SPECIES 1.

Madrepora turbinata. *Linn.*

Conical, longitudinally striated; upper part concave, radiated; radii numerous, unequal in length.

Farley.

AGGREGATED.

SPECIES 2.

Conical, covered by slightly concave, closely radiated stars.

Stoke.

SPECIES 3.

Large, depressed, concave and patulous above; stars concave, small, numerous; base closely striated from the pedicle.

Castle Combe.

SPECIES 4.

Large, convex above, with many equal concave stars.

a Castle Combe.

b Combe Down.

SPECIES 5.

Large, convex above, covered with many small concave stars; surface undulated.

Vinyard Down.

FASCICULATED.

SPECIES 6.

Madrepora flexuosa. *Linn.*

Spreading; branches round, longitudinally striated, rough with transverse wrinkles, concave at the upper end, with radii unequal in length.

Castle Combe.

SPECIES 7?

Spreading at the sides, composed of small round striated branches.

This may perhaps be a Tubipora.

a Broadfield Farm.

b Combe Down.

SPECIES 8.

Part of a large cylindrical ramose Coral.

Castle Combe.

TESTACEA.

BIVALVIA.

INEQUIVALVED BIVALVES.

PECTEN.

SPECIES 1.

Subequilateral, convex, beak central; many close longitudinal ribs. Length more than two inches.

Castle Combe.

SPECIES 2.

Pecten, Species 1. *Forest Marble.*

Large, convex, eared, ribbed longitudinally with forty or fifty wide smooth ribs; striated transversely, striæ obscure on the ribs.

Length two inches or more.

Cotswold Hills.

PLAGIOSTOMA.

Plagiostoma, Species 2. *Clay over the upper oolite.*

VARIETY. More depressed.

Length one inch and a half.

Bradford?

OSTREA.

Ostrea crista galli, Cock's comb oyster.

Petty France.

TEREBRATULA.

NOT PLICATED.

SPECIES 1.

Terebratula, Species 1. *Clay over the upper oolite.*

Obscurely five sided, rather depressed, smooth; larger valve most convex; front margin undulated, with three depressions in the smaller valve and four in the larger.

Petty France.

PLICATED.

SPECIES 2.

Terebratula obsoleta? *Min. Conch.*

Nearly round, gibbous, plaited; middle of the front a little elevated, with from six to eight plaits; sides having from nine to twelve plaits; beak projecting.

Petty France.

ECHINI.

ANOCYSTI.

CLYPEUS.

Clypeus. *Coral Rag and Pisolite.*

Hinton.

Hogwood Corner.

FULLER'S EARTH ROCK.

TESTACEA.

UNIVALVIA.

TROCHUS.

Cast of the inside of a Trochus with roundish volutions.
Charlton Horethorn.

TUBULAR IRREGULAR UNIVALVES.

SERPULA.

SPECIES 1.

Elongated, conical, twisting at the small end; quadrate with four projecting crenated angles or ridges; aperture round.

a Orchardleigh.

b Charlton Horethorn.

SPECIES 2.

Serpula crassa. *Min. Conch.*

Acutely conical, round within, three-sided externally, about four times as long as the diameter of the end at the aperture.

Charlton Horethorn.

MULTILOCULAR UNIVALVES.

NAUTILUS.

Fig. 1, *Fuller's Earth Rock Plate.*

Flatted globose, umbilicate; shell with transverse ridges curving from the aperture; aperture wider than long, siphunculus central.

It seems as if the transverse ridges are formed by the endings of thick plates, and at first sight they resemble septa, but curve from the aperture.

Lansdown.

BELEMNITES.

Slender, with a deep groove from the apex.

Charlton Horethorn.

AMMONITES.

SPECIES 1.

Ammonites modiolaris.

Fig. 2, Fuller's Earth Rock Plate.

Subglobose, inner volutions convex, exposed within a large conical hollow or umbilicus; back wide; aperture semilunate, rounded at the extremities, three times as wide as long.

The inner volutions are convex and sharply undulated within the umbilicus, with many large furrows across the back curving toward the aperture; the outer volutions are irregularly swelled on the inner edge, and the undulations across the back are obscure. The umbilicus is one third of the greatest diameter.

This species may be distinguished from Ammonites sublævis by the large size of the umbilicus, and the obtuseness of the angle which bounds it.

Dundry.

Rowley Bottom.

SPECIES 2.

Subglobose, convex, deeply umbilicate; umbilicus narrow, undulated, with a sharpish margin; transverse radiating undulations across the back uniting on the inner part: aperture semilunate or deltoidal, truncated at the sides, the width two and a half times as great as the length.

Hardington?

Broadfield Farm.

BIVALVIA.

EQUIVALVED BIVALVES.

MODIOLA.

SPECIES 1.

Coated muscle.

Near Bath.

SPECIES 2.

Elongated, slender, sides subparallel, one side thinned and projecting; flattish, lines of growth distinct.

Ancliff.

SPECIES 3.

Long, with subparallel sides; gibbous diagonally from the beaks.

Near Bath. A large cast of the inside.

SPECIES 4.

Modiola anatina.

Fig. 3, Fuller's Earth Rock Plate.

Oblong, very gibbous, smooth; beaks prominent, hooked, subinvolute; one side arched into the front; the other oval, flat or concave with a very thick blunt projection near the beaks; shell with fine striæ or lines of growth.

Length twice the width, thickness two thirds of the length.

In a side view the shell is regularly oval with a notch between the beaks.

Ancliff.

UNIO.

SPECIES 1.

Transverse, elongated, depressed, transversely furrowed, beaks pointed, nearest to one end; front margin almost straight; length not half the breadth; thickness half the length.

a Grip Wood. Cast of the inside.

b Hardington. Cast of the inside.

SPECIES 2.

More than twice and a half as wide as long, transversely striated, front straight, sides acute, posterior side swelled near the edge; beaks nearest to one end.

a Turnpike near Bratton.

b Grip Wood.

SPECIES 3.

Triangular, wider than long, smooth, sides acute; anterior side projecting; beaks hooked above the posterior flat side; front margin concave near the anterior angle.

Grip Wood?

ASTARTE.

Cast of the inside of Astarte ovata.

Grip Wood.

TRIGONIA.

SPECIES 1.

Trigonia clavellata. Small.

Orchardleigh.

SPECIES 2.

Cast of a large species ; cuneiform, very thick, with two muscular impressions.

Breadth three inches and a quarter.

Devonshire Buildings, Bath.

SPECIES 3.

Cast of the inside ; transversely elongated, cuneiform, very gibbous : one end pointed, two muscular impressions.

Grip Wood.

SPECIES 4.

Cast of the inside ; transversely elongated, pointed ; beaks close to the short rounded side ; breadth twice the length.

CARDITA.

SPECIES 1.

Fig. 4, *Fuller's Earth Rock Plate.*

Large, rhomboidal, very gibbous, beaks at one of the angles, hooked ; surface covered with obscure curvilinear transverse furrows and a few longitudinal ones.

a Grip Wood.

b Hardington.

SPECIES 2.

Deltoidal, very gibbous, beaks subinvolutè, curved over the posterior flat side ; transversely striated; thickness equal to the length.

Ancliff.

SPECIES 3.

Cast of the inside of a small species ; circular, gibbous, margin crenulated within.

Near Bath.

CARDIUM.

Cardium. *Cornbrash.*

Fig. 5, *Fuller's Earth Rock Plate.*

Subglobose, ovate, anterior slope straight ; surface undulated with about ten large ridges, crossed by transverse furrows ; beaks at one end, very protuberant, incurved.

Length two or three inches.

Specimens from this stratum are generally wider and less globose than in the Cornbrash ; generally a large prominent ridge bounding the posterior side. The longitudinal ridges are widened and flattened in the anterior part.

a Charlton Horethorn.
b Near Gagenwell.
c Near Redlynch.

MACTRA. *Linn.*

SPECIES 1.

Transversely subovate, or four sided, gibbous, subequilateral; beaks prominent, front straight, sides truncated, gaping, transversely furrowed.

Grip Wood.

SPECIES 2.

Transversely elongated, arched, depressed; covered with longitudinal furrows; posterior side rounded; anterior side flattened, a little gaping, not expanded. Breadth twice the length and three times the depth.

The longitudinal furrows are decussated by distant broad transverse undulations.

Cotswold Hills.

SPECIES 3.

Mactra gibbosa. *Min. Conch.*

Gibbose, anterior side much wider than the posterior, recurved, truncated, gaping: length and depth about equal, breadth equal to twice the length: posterior side rounding.

Mitford, a very large specimen, six inches wide.

VENUS?

SPECIES 1.

Venus, Species 1. *Cornbrash.*

VARIETY. Wider.

Near Gagenwell.

SPECIES 2.

Circular, flattish; beak prominent.

Charlton Horethorn. Cast of the inside.

SPECIES 3.

Oval, wider than long, thickish, transversely sulcated.

Bath. Cast of the inside.

TELLINA.

Fig. 6, *Fuller's Earth Rock Plate.*

Transverse, elongated, one side extended into a long beak, the other flattish and circular at the extremity; transversely furrowed.

The back is arched from the posterior or circular end, the elongated side straight and hollowed under the beaks: a wide depression from the beaks near the posterior end; width three or four times the length.

a Ancliff.

b Hardington.

MYA?

SPECIES 1.

Transverse, ovate, beaks close to the posterior side; anterior side acute; transversely furrowed.

a Dundry.

Sherborn.

SPECIES 2.

Mya intersectans.

Depressed, wide, anterior side expanded; covered by intersecting furrows, longitudinal on the anterior part, oblique on the posterior part, meeting in an oblique line: beaks prominent, nearest to the posterior end.

INEQUIVALVED BIVALVES.

OSTREA.

SPECIES 1.

Ostrea Marshii. *Min. Conch.*

Fig. 8, *Fuller's Earth Rock Plate.*

Oblique, both valves deeply plaited in seven or eight angular diverging undulations; edge thick, flatted.

a Monkton Combe.

b Cotswold Hills. This has fine longitudinal striæ.

SPECIES 2.

Ostrea rugosa.

Valves equal, oblong, depressed, one side eared; very rough, with eight or more broad longitudinal ridges elevated into projecting spinous tubercles: hinge square, flat, pit hollowed, with straight parallel edges, the posterior end curving to the beak.

The shell is uncommonly thick; the hinge very massive, and projecting far into the shell; one of the flat lateral parts forms a rectangular ear.

a Monkton Combe.

b Between Nunny and Frome.

SPECIES 3.

Circular, smooth.

Orchardleigh, upper valve, small.

TEREBRATULA.

NOT PLICATED.

SPECIES 1.

Terebratula ornithocephala. *Min. Conch.*

Ovato-rhomboidal; depressed when young, elongated and gibbous when old; front straight, bounded by two obtuse lateral depressions, similar in each valve.

a Turnpike near Bratton.

b Near Bath.

VARIETY.

Very long, gibbous, nearly as thick as wide; valves subcarinated along the middle; front narrow, beaks much elongated and curved.

a Near Bath.

b Writhlington.

c Turnpike near Bratton.

SPECIES 2.

Terebratula intermedia. *Min. Conch.*

Some of the specimens resemble Terebratula biplicata.

Near Bath.

PLICATED.

SPECIES 3.

Terebratula media. *Min. Conch.*

Fig. 9, *Fuller's Earth Rock Plate.*

Very obtusely deltoid, gibbous, plaited; front rounded, with a rising in the middle, composed of six sharp plaits approaching those in the sides; beak a little incurved.

a Near Bath.

b Charlton Horethorn.

c Orchardleigh.

ECHINI.

CATOCYSTI.

CONULUS.

Nearly circular, convexo conical, covered by small prominences, particularly on the bas of the shell; rays ten biporous lines diverging in five pairs; pores very close on the uppe part, more distant on the base, which is concave.

Canal at Bradford.

UNDER OOLITE.

ZOOPHITA.

MADREPORA.

SIMPLE, COMPOSED OF ONE STAR.

SPECIES 1.

Madrepora porpites. *Linn.*
a Bath.
b Churchill.

SPECIES 2.

Circular, discoidal; upper side flat, with many radii unequal in length, lower side with two or three circular furrows, closely striated from the center.
Bath.

AGGREGATED.

SPECIES 3.

Stars oblong, very unequal, irregular, radii unequal in length, diverging in pencils or small bundles.
a Dundry.
b Bath Hampton.

SPECIES 4.

Stars pentagonal or hexagonal, deeply hollowed, radiate, septa distinct, stars small, regular.
Dundry.
Tucking Mill.
Crickley Hill.

SPECIES 5.

Columnar, reticulated.
Bath Hampton.

A CORAL.

Composed of cylindrical ramose striated branches.
Bath.

PENTACRINUS.

Small, angles acute, very projecting.
Mitford. Mass.

TESTACEA.

UNIVALVIA.

CONUS.

Conical, spire short, pointed, one sixth of the length; aperture narrow. Length an inch and a half.

Near Bath. A cast of the inside.

MELANIA.

SPECIES 1.

Small, elongated, aperture roundish, oblong; spire set with oblong tubercles.
Tucking Mill. A cast of the inside.

SPECIES 2.

Casts of a Melania, four or five times as long as the diameter.
a Coal Canal.
b Tucking Mill.
c Near Bath.

SPECIES 3.

Melania striata. *Min. Conch.*
Caisson.

TROCHUS.

SPECIES 1.

Conical, high; volutions seven or eight, the lower part swelling into a prominent crenated rim, the upper edge of one volution touching the rim of the next; shell reticulato-striated; a few longitudinal wrinkles on the upper part of the volution; base flat, no umbilicus.

Near Bath.

SPECIES 2.

Conical, high; volutions five or six, even, with a crenulated rim on the lower edge.
a Coal Canal. Cast of the inside.
b Mitford Inn. Cast of the inside.

SPECIES 3.

Conical, volutions angular, increasing quickly, flat in the upper part; base flattish, columella perforate.

Coal Canal. Cast of the inside.

SPECIES 4.

Conical, sides even, volutions five or six, base convex; columella perforate.

Coal Canal. Cast of the inside.

SPECIES 5.

Short conical, volutions four, the upper ones roundish, the lower ones angular, with a projecting upper edge.

a Coal Canal. Cast of the inside.

b Tucking Mill. Cast of the inside.

SPECIES 6.

Conical, volutions few, roundish.

a Coal Canal. Cast of the inside.

b Tucking Mill. Cast of the inside.

c Between Cross Hands and Petty France. Cast of the inside.

SPECIES 7.

Depressed, very short, volutions three or four, opening roundish or ovate, wider than long.

a Coal Canal. Cast of the inside.

b Tucking Mill. Cast of the inside.

SPECIES 8.

Short conical, depressed, volutions three, angular toward the lower part, base discoidal, convex.

a Coal Canal.

b Near Bath.

SPECIES 9.

Depressed, volutions four or five, with two small rims on the lower part, the upper one nearly smooth, the lower one crenated; shell reticulato-striated; base convex, without an umbilicus; opening roundish.

Sherborn.

Bath.

PLANORBIS.

Flat above ; volutions few, angular, roundish on the inside, widest above.

a Tucking Mill. Cast of the inside.

b Near Bath. Cast of the inside.

TURRITELLA.

SPECIES 1.

Elongated, volutions numerous, with a deep spiral groove.

a Smallcombe Bottom. Cast of the inside.

b Coal Canal. Cast of the inside.

c Tucking Mill. Cast of the inside.

SPECIES 2.

Very much elongated, slender; one distinct spiral projection in the upper part and many spiral striæ : aperture oblong, narrow. In one cast there appear three projections.

Near Bath.

SPECIES 3.

Turreted with many small costæ, acute on the upper edge.

a Churchill. VARIETY. With spiral striæ.

b Near Bath.

AMPULLARIA.

SPECIES 1.

Very small, subglobose, spire very short, aperture semicircular.

Crickley Hill. Cast of the inside.

SPECIES 2.

Volutions three or four, the outer one large, spire short, with an obtuse furrow beneath the upper edge of the volutions : aperture semicircular.

Length about an inch and a half.

a Coal Canal. Cast of the inside.

b Bath. Cast of the inside.

SPECIES 3.

Ovate, spire acute, volutions three or more, convex, smooth. Aperture half the length of the shell.

Length not more than one inch.

a Tucking Mill. Cast of the inside.

b Coal Canal. Cast of the inside.

c Near Bath. Cast of the inside.

TURBO.

Volutions about six, with several muricated lines; one at the upper edge, another projecting about the middle of the volution, and five or more beneath it; many longitudinal striæ crossing from one row to another: mouth rather oblong, base of the columella acute.

Length from half to one inch.

The tubercles are sharp and very regular, the cross striæ are about three to each tubercle.

a Tucking Mill.

b Near Bath.

Sherborn.

TUBULAR IRREGULAR UNIVALVES.

SERPULA.

SPECIES 1.

Cylindrical, twisting; diameter one fourth of an inch.

Near Bath.

SPECIES 2.

Elongated, conical, twisting at the small end, quadrate with four projecting crenated angles or ridges; aperture round.

Churchill.

MULTILOCULAR UNIVALVES.

NAUTILUS.

SPECIES 1.

Volutions partly visible in the cast, back round; aperture ovato-lunate, its breadth twice its length; septa many, siphuncle nearest to the outside.

Depth three fourths of the longest diameter, which in one specimen is six inches and a half.

Sherborn.

SPECIES 2.

Convex, back flatted, inner volutions concealed in the cast; siphuncle central; mouth lunate, narrow at the extremities; length two thirds of the breadth.

Longest diameter four inches and a half.

Sherborn.

SPECIES 3.

Subglobose, umbilicate, back very wide, volutions few; siphuncle central; aperture twice as broad as long.

Shell longitudinally striated ; lines of growth distinct.

a Sherborn.

b Between Sherborn and Yeovil.

c Charlton Horethorn.

BELEMNITES.

SPECIES 1.

Slender, with a deep groove from the apex.

Tucking Mill.

Sherborn.

SPECIES 2.

Long, slender, without any groove.

a Yeovil.

b Wotton Underedge.

AMMONITES.

NO KEEL OR FURROW ON THE BACK.

SPECIES 1.

Ammonites modiolaris.

Subglobose, inner volutions convex, exposed within a large conical hollow or umbilicus ; back wide ; aperture semilunate, rounded at the extremities, three times as wide as long.

Lansdown.

SPECIES 2.

Ammonites calix.

Discoidal, concave or cup formed, radiated ; back wide, slightly convex ; radii prominent, rather distant, terminating in tubercles, and forming a deeply indented margin to the cup ; tubercles dividing over the back into three transverse ribs : aperture much wider than long.

The longest diameter in one specimen is five inches.

The transverse ribs are in number about three to each tubercle, on the outer volutions four, or more. The aperture is a portion of a circular ring bounded by two radiating lines.

Sherborn.

SPECIES 3.

Subglobose when young, more depressed when old, umbilicate ; margin of the umbilicus rounded ; very numerous small radii across the back, united on the inner part of the volution, undulating the umbilicus ; aperture three times as wide as long, semicircular, the sides rounded into the umbilicus.

Sherborn.

SPECIES 4.

Depressed, convex, volutions seven, roundish on the back, the inner ones half concealed; radii numerous, large on the inner part, smoothed along the middle, very obscurely dividing into four or five, which are obsolete on the middle of the back; aperture oval, one third of the diameter long, inner edges obtuse.

Longest diameter seven inches.

Sherborn.

SPECIES 5.

Volutions five or six, inner ones exposed; concave in the middle; radii numerous, sharp, bifurcating over the wide back; aperture about two-fifths of the diameter in length and as wide as long.

Sherborn.

Dundry.

SPECIES 6.

Ammonites communis. *Min. Conch.*

Sherborn.

A KEEL ON THE BACK.

SPECIES 7.

Ammonites radiatus.

Discoidal, very flat, concave in the middle, outer edge acuminated, with a swelling undulated keel; volutions very quickly increasing, leaving a central hollow; surface undulated with broad radiating undulations and many concentrate spreading furrows.

The shell is very thin, with radiating and concentrate striæ. Aperture sagittate, inner angles circular. Septa near together, very much undulated.

Longest diameter almost eight inches.

Between Sherborn and Yeovil.

SPECIES 8.

Small, discoidal, concave in the middle; volutions four, half concealed; radii obsolete; aperture oval; keel small, sharp.

Near Bath.

SPECIES 9.

Ammonites concavus. *Min. Conch.*

Involute, depressed, keeled, umbilicate; umbilicus a large hemispherical depression; volutions four, depressed near the center; radii numerous, curved, unequal in length, obsolete near the center; keel sharp, entire; aperture acutely triangular, external angle rounded, internal angles obliquely truncate.

Variety. With obtuse radii, and each side of the back bevelled.
Between Sherborn and Yeovil.
Sherborn.
Dunkerton.

A KEEL BETWEEN TWO FURROWS ON THE BACK.

SPECIES 10.

Volutions four or five, almost half concealed; back wide, with a high keel and a slight furrow on each side; radii obtuse, unequal in length, some of them furcate, curved a little on the edges of the back; aperture oblong, its length two fifths of the diameter.
Near Bath.

A FURROW ON THE BACK.

SPECIES 11.

Volutions five or six, gibbous, inner ones a third concealed, back rounded; radii many, sharp, bifurcate with a few distant alternating entire ones, acute on the edges of a flat or concave space along the back: aperture obcordate.
Near Bath.

BIVALVIA.

EQUIVALVED BIVALVES.

MODIOLA.

SPECIES 1.

Coated Muscle. Fistulana. *Lam.*
a Bath.
b Mitford. Incrusted with coral.
Tucking Mill.

SPECIES 2.

Very much elongated, with parallel sides, narrowing at the extremities, gibbous diagonally from the beaks; back margin with oblique short furrows; lines of growth distinct: width four or five times the length.
Churchill.

SPECIES 3.

Very large, twice as broad as long.
Oldford near Frome. Cast of the inside.

SPECIES 4.

Modiola. *Cornbrash.*
Churchill.
Northwest of Northampton.

SPECIES 5.

Triangular, depressed, gibbous diagonally from the beaks; posterior side small, anterior side expanded, straight, digonal; back and front straight; beaks small, close to the posterior end; greatest width almost three inches; greatest length, at the anterior end, above half the width.

Dundry.

UNIO.

SPECIES 1.

Subtriangular, broader than long, depressed, transversely striated; front margin straightish; an inflection on one side.

Between Sherborn and Yeovil.

SPECIES 2.

Wider than long, depressed, posterior side elliptical, arched round to the beaks, anterior side shorter, hollowed under the beaks.

Churchill.

SPECIES 3.

Unio, Species 3. *Fuller's Earth Rock.*
Northwest of Northampton.

ARCA.

Wider than long, very gibbous, straightish at the edges, with sharp transverse lines of growth, and a few longitudinal striæ on the sides; beaks distant; shell thick.

Sherborn.

Dundry. A very large cast of the inside.

TRIGONIA.

SPECIES 1.

Trigonia costata. *Park. Min. Conch.*

Triangular, with transverse smooth ribs; anterior side marked with many small and three large prominent longitudinal crenulated ridges.

Posterior angle very obtuse; anterior side large; the transverse ribs terminate before the first longitudinal ridge bounding the anterior side.

a Cotswold Hills. Cast of the inside.
b Between Cross Hands and Petty France. Cast of the inside.
c Cross Hands. Cast of the inside.
d Mitford. Cast of the inside.
e Coal Canal. Cast of the inside.
f Tucking Mill.
Little Sodbury.

SPECIES 2.

Cast of the inside; depressed, wider than long.
a Crickley Hill.
b Nailsworth.

INSIDE CASTS OF BIVALVES,

WITH TWO MUSCULAR IMPRESSIONS IN EACH VALVE.

SPECIES 1.

Rhomboidal, gibbous in the middle, flattish at the sides; muscular impressions large, convex in the cast, furrowed; beaks of the cast distant. Length two inches or more.
a Bath.
b Mells.

SPECIES 2.

Circular, depressed, with projecting beaks; margin thin; muscular impressions convex in the cast, furrowed; beaks of the cast distant.
Bath.

CARDIUM.

Cardium. *Cornbrash.*
a Writhlington.
b Between Sherborn and Yeovil.
c Chipping Norton.
d Churchill.
e Dundry.
Northwest of Northampton.

MACTRA.

SPECIES 1.

Mactra, Species 2. *Fuller's Earth Rock.*

Transversely elongated, arched, depressed, covered with longitudinal furrows; posterior side rounded, anterior side flattened, a little gaping, not expanded; breadth twice the length, and three times the depth.

The longitudinal furrows are decussated by distant broad transverse undulations.

a Crewkerne:

b Churchill.

SPECIES 2.

Mactra gibbosa. *Min. Conch.*

Gibbose, anterior side much wider than the posterior, recurved, truncated, gaping: Length and depth about equal, breadth equal to twice the length; posterior side rounding.

a Tucking Mill. A very large specimen, six inches wide.

b Mitford.

SPECIES 3.

Large, squarish, subequilateral, gibbous, with contiguous prominent beaks; nearly straight at the edges, transversely undulated. Length three fourths of the breadth, depth two thirds of the breadth, which is about three inches.

Gloucestershire.

ASTARTE.

Astarte ovata. *Oaktree Clay.*

Transversely oblong, depressed, anterior side lengthened, transversely striated; lunette elliptical; margin crenulated within; shell thick; beak acute, solid: a small pit beneath the posterior slope of the hinge.

a Between Sherborn and Yeovil.

b Coal Canal. Cast of the inside.

c Tucking Mill. Cast of the inside.

d Bath. Cast of the inside.

e Fulbrook. Cast of the inside.

f Between Cross Hands and Petty France. Cast of the inside.

Mitford Inn. Cast of the inside.

Northwest of Northampton. A very small Cast of the inside.

MYA.

Mya, Species 1. *Fuller's Earth Rock.*

Transverse, ovate, beaks close to the posterior side; anterior side acute; transversely furrowed.

Dundry.

INEQUIVALVED BIVALVES.

PLAGIOSTOMA.

A small variety of Plagiostoma gigantea. *Min. Conch.* with very distinct longitudinal striæ.

Tucking Mill.

LIMA.

SPECIES 1.

Lima gibbosa. *Min. Conch.*

Elongated, gibbose, smooth, longitudinally plicated in the middle; ears undefined; nearly twice as long as wide. About eighteen small sharp plaits in the middle.

Churchill.

SPECIES 2.

Trigonal, oblique, with short ears; one side straightish and lengthened, the other rounded, covered with many longitudinal sharp plaits.

Churchill.

CRENATULA.

Flat, or much depressed, oblong, sides subparallel, widest near the beaks, which are at one end of the hinge line.

Dundry.

OSTREA.

SPECIES 1.

Ostrea rugosa. Species 2. *Fuller's Earth Rock.*

a Tucking Mill.

b Between Sherborn and Yeovil. This is smaller, more regular and thinner.

SPECIES 2.

Oblong, straight, sides subparallel; depressed; lower valve with transverse imbricated waves; upper valve flat, transversely furrowed.

Chipping Norton.

Northwest of Northampton.

SPECIES 3.

Ostrea acuminata. *Min. Conch.*

Churchill.

SPECIES 4.

Small, oblong, lobate.

Betwen Cross Hands and Petty France. Much distorted.

AVICULA.

SPECIES 1.

Avicula costata. VARIETY, with many ribs.
a Between Cross Hands and Petty France.
b Tucking Mill.

SPECIES 2.

Similar to avicula costata, but the ribs are tuberculated.
Between Sherborn and Yeovil.

PECTEN.

SPECIES 1.

Pecten equivalvis. *Min. Conch.*

Lenticular, with rounded diverging ribs and many acute concentric striæ; valves equally convex, the lower one smoothest; ears equal.

a Ilmington.
b Dursley.
Dowdswell Hill.
Sherborn.

SPECIES 2.

Small, flattish, smooth.
Churchill.

SPECIES 3.

Pecten fibrosus. *Min. Conch.*
Churchill.

INOCERAMUS.

Fibrous Shell.

Between Cross Hands and Petty France.
Monkton Combe.
Tucking Mill.
Bath.

TEREBRATULA.

NOT PLICATED.

SPECIES 1.

Terebratula intermedia? *Min. Conch.*
a Batheaston.
b Near Lansdown.
c Tucking Mill.
d Between Sherborn and Yeovil.
e Churchill.
f Chipping Norton.
g Fulbrook.

SPECIES 2.

Terebratula ornithocephala. *Min. Conch.*
Variety. The edges flattened.
Tucking Mill.
Sherborn.

PLICATED.

SPECIES 3.

Terebratula obsoleta? Species 3. *Clay over the upper Oolite.*
a Between Cross Hands and Petty France.
b Tucking Mill.
c Churchill.
d Chipping Norton.
e Fulbrook.

SPECIES 4.

Terebratula spinosa.

Circular, convex, with many roundish plaits set with long slender spines in quincuncial order; middle of the front elevated in a large wave rounding into the sides; beak small, incurved; upper valve most convex.

a Bath.
b Tucking Mill.
Chipping Norton.

ECHINI.

ANOCYSTI.

CIDARIS.

SPECIES 1.

Cidaris, Species 1. *Clay over the upper oolite.*
Tucking Mill.

SPECIES 2.

Pentangular, depressed, with projecting rather distant small mammellæ; two contiguous rows in each areola, and four converging rows in each area; the two middle rows short and only on the side or widest part of the area; rough with small points encircling the mammellæ; rays obliquely triporous.

The areolæ form the angles of the pentagon. The two longer rows of mammellæ in each area are parallel to the rays and converge to the aperture, and the space between them on the side is occupied by two shorter converging rows.

Tucking Mill.

SPECIES 3.

Cidaris, Species 3. *Coral Rag and Pisolite.*
Tucking Mill.

CLYPEUS.

SPECIES 1.

Clypeus sinuatus. *Leske.*

Circular, depressed, convex above, base rather concave, with five diverging furrows; rays ten, in five pairs, approaching each other on the circumference, transversely striated with a line of pores on their margin; one of the areæ divided by a deep sharp furrow from the vertex; mouth small, pentagonal; shell covered by small eminences in circular depressions.

a Monkton Combe.
b Stunsfield.
c Chipping Norton.
d Churchill.
e Fulbrook.
f Near Naunton.
g Stow on the Wold.
Northwest of Northampton.

SPECIES 2.

Clypeus. *Coral Rag and Pisolite.*
Churchill.

CATOCYSTI.

CONULUS.

Conulus. *Fuller's Earth Rock.*
Tucking Mill.

Reference to the figures in the Plates of this Stratum, which will be given in "Strata identified by Organized Fossils," not having been regularly inserted, it is necessary to observe that the following species will be figured.

First Plate.

Melania,	Species 2	Fig. 1
Trochus,	Species 1	2
	Species 6	3
	Species 9	4
Turritella,	Species 1	5
	Species 3	6
Ampullaria,	Species 2	7
Nautilus,	Species 3	8
Ammonites	calix	9

Second Plate.

Madrepora, Species 4	Fig. 1
Trigonia costata	2–3
Astarte ovata	4
Pecten equivalvis	5
Inoceramus	6
Terebratula spinosa	7
Clypeus sinuatus	8

SAND AND SANDSTONE.

TESTACEA.

UNIVALVIA.

MULTILOCULAR UNIVALVES.

BELEMNITES.

Fig. 1, *Sand and Sandstone Plate.*
Tapering, apex of the alveolus central. Diameter two thirds of an inch.
Tucking Mill.

AMMONITES.

NO KEEL OR FURROW ON THE BACK.

SPECIES 1.

Ammonites, Species 4. *Marlstone.*
Volutions about four, thick, the inner ones not much concealed, round on the back; radii very numerous, bifurcating at the middle; aperture round, lunate on the inner side.
Enstone.

A KEEL ON THE BACK.

SPECIES 2.

Ammonites ellipticus? *Min. Conch.*
Yeovil.

SPECIES 3.

Ammonites, Species 8. *Marlstone.*
In this specimen, the radii are not so much bent in the middle as those in the marlstone.
a Liliput.
b Tucking Mill.

SPECIES 4.

Fig. 2, *Sand and Sandstone Plate.*
Much depressed, volutions few, half exposed, quickly increasing, thin at the outer edge; radii small, numerous, twice rather obtusely curved, unequal in length, uniting by pairs or

more into short prominences on the inner part: keel sharp; aperture two fifths of the longest diameter.

Yeovil.

BIVALVIA.

EQUIVALVED BIVALVES.

MODIOLA.

Fig. 3, *Sand and Sandstone Plate.*

Very much elongated with nearly parallel sides, acuminated at the extremities, gibbous diagonally from the beaks; posterior side, or hinge side, furrowed; lines of growth distinct, parallel to the margin. Length four or five times the breadth. The furrows on the back margin are oblique, short, and do not reach to the gibbous part.

Top of Frocester Hill.

MYA.

Mya, Species 1. *Fuller's Earth Rock.*

Enstone.

INEQUIVALVED BIVALVES.

AVICULA.

Avicula costata.

Enstone.

TEREBRATULA.

Terebratula ornithocephala. *Min. Conch.*

Enstone. A curious variety with flattened edges, similar to those from Tucking Mill and Sherborn in the under oolite.

MARLSTONE.

ZOOPHITA.

MADREPORA.

SPECIES 1.

Madrepora porpites. *Linn.*
Tucking Mill.

PENTACRINUS.

SPECIES 1.

Fig. 1, *Marlstone Plate.*
Vertebræ thin, acute in the middle with five very prominent angles; vertebral column excavated at the meeting of the vertebræ between the prominent angles.
Churchill.
Stone Farm, Yeovil.
Kennet and Avon Canal.

SPECIES 2.

Vertebræ pentagonal, sides flat.
Wotton under Edge.

TESTACEA.

UNIVALVIA.

HELIX.

Helix. *Coral Rag and Pisolite.*
Wotton under Edge.

MULTILOCULAR UNIVALVES.

BELEMNITES.

SPECIES 1.

Large, squarish, with two obtuse furrows very near the apex.
Penard Hill.

An Alveolus of a large belemnite; very acutely conical, septa distant from each other one sixth of their diameter.

Foot of Frocester Hill.

Alveolus of a large belemnite; long, conical, septa numerous, distant from each other one seventh of their diameter.

Tucking Mill.

Long, slender.

Fig. 2, *Marlstone Plate.*

a Yeovil.

b Churchill.

c Tucking Mill.

d Enstone.

NAUTILUS.

A depressed umbilicated Nautilus, resembling Nautilus lineatus, *Min. Conch.* is found in this Stratum, with crystallized Carbonate of Lime in the chambers; the shell is striated longitudinally and appears to be very thin; the situation of the siphuncle has not been observed in the specimen.

Tucking Mill.

AMMONITES.

NO KEEL OR FURROW ON THE BACK.

SPECIES 1.

Depressed, volutions four or five, the back smooth, nearly flat, thicker than the other parts of the volution; radii distant, acute on the outer margin; aperture quadrangular, widest in the outer part.

Coal Canal.

SPECIES 2.

Depressed, volutions five or more, the inner ones more than a fourth concealed, with many obtuse distant radii smoothed over the back; aperture elliptical, indented, its length a fourth of the diameter.

The radii are most prominent on the middle of the volution.

Coal Canal.

SPECIES 3.

Ammonites undulatus.

Fig. 3, *Marlstone Plate.*

Depressed, radii obtuse, mostly equal in length, twice curved, undulating the back; volutions four or more, depressed along the middle, the inner ones three fourths concealed: aperture very oblong, inner edges truncated.

The radii are obscure towards the center, twice curved, and on the back incline toward the aperture, and are so prominent as to give it the appearance of being folded or plaited. In some specimens the radii seem closer and more numerous on the outer volutions than on the inner.

Coal Canal.

SPECIES 4.

Fig. 4, *Marlstone Plate.*

Volutions about four, thick, the inner ones not much concealed, round on the back; radii very numerous, bifurcating at the middle; aperture round, lunate on the inner side. Diameter three fourths of an inch to two inches and a half.

a Coal Canal.
b Tucking Mill.
c Yeovil.
d Churchill.

SPECIES 5.

Ammonites communis. *Min. Conch.*

This shell varies much. These specimens are of the variety with an oblong aperture.

a Near Bath.
b Tucking Mill.
c Coal Canal.
d Dundry.

SPECIES 6.

Large, volutions three, very quickly diminishing in depth and diameter, not at all concealed; radii, numerous irregular undulations, most distinct on the back, which is round; opening oval, the inner angles blunt, two fifths of the longest diameter. Diameter seven inches and a half.

Yeovil.

A KEEL ON THE BACK.

SPECIES 7.

Volutions few, quickly increasing, the inner ones more than half exposed, blunt at the back, inner edges truncate and smooth; radii very numerous, most prominent on the outer part of the volution, twice curved, the first bend in a slight spiral depression near the middle; keel small, acute; aperture narrow, more than two fifths as long as the diameter. Longest diameter six inches.

This is probably a different species from Ammonites elegans, *Min. Conch.* to which, in many respects, it is very similar; the most essential distinctions are the bluntness or width of the back, and th egreater exposure of the inner volutions.

a Yeovil.

b Tucking Mill.

SPECIES 8.

Volutions few, the inner ones two fifths concealed, obtuse at the back, truncate at the inner angles; radii large, obscure on the inner half of the volution, conspicuous on the outer half, twice curved, the first bend in a spiral depression near the middle; aperture narrow, with subparallel sides, inner angles truncate. Diameter five inches.

This species differs from the preceding in the lesser number and greater size of its radii; in other respects perfectly similar.

a Stone Farm, Yeovil.

b Penard Hill.

SPECIES 9.

Ammonites concavus. *Min. Conch.*

Coal Canal.

SPECIES 10.

Fig. 5, *Marlstone Plate.*

Depressed, volutions three or four, the inner ones almost half concealed; radii furcating from the inner part, bent in the middle, curved toward the keel in the outer part; back flattish, keel small, acute: aperture narrow, inner angles truncate. Diameter between one and two inches.

a Coal Canal.

b Tucking Mill.

c Penard Hill.

Dundry.

Frocester Hill.

Bathhampton, foot of inclined plane.

SPECIES 11.

Depressed, volutions four or five, convex, the inner ones a third concealed; radii many, obtuse, curving, unequal in length; septa rather distant, equal in number to the entire radii; aperture oval, a third as long as the diameter, without any truncation on the inner angles. Diameter about two inches.

Glastonbury.

A KEEL BETWEEN TWO FURROWS ON THE BACK.

SPECIES 12.

Volutions four, gibbous, almost wholly exposed; radii many, a little curved, undulating the inner margin; aperture squarish, with rounded angles.

Coal Canal.

SPECIES 13.

Ammonites Walcotii. *Min. Conch.*

Fig. 6, *Marlstone Plate.*

Involute, depressed; volutions four, three fourths exposed, with a concentrate furrow; lunate undulations over half the sides; back with a keel between two furrows; aperture oblong, one third of the diameter long.

Each volution divided into two parts by an obtuse furrow; inner half nearly smooth.

Coal Canal.

Glastonbury.

A FURROW ON THE BACK.

SPECIES 14.

Much depressed, radii entire, sharp, prominent and opposite on the outer margins; opening oblong, narrow, sides almost parallel, back concave along the middle.

Near Bath.

BIVALVIA.

INEQUIVALVED BIVALVES.

OSTREA.

Gryphea, Species 1. *Lias.*

Northeast of Newark.

PECTEN.

Fig. 7, *Marlstone Plate.*

Circular, depressed with many longitudinal obtuse ribs.

Kennet and Avon Canal. A mass.

Northeast of Newark.

TEREBRATULA.

Terebratula obsoleta. *Min. Conch.*
a Churchill.
b Wotton under Edge.

The organized Fossils in the under part of the rock differ very materially from those in the upper, but these distinctions, with others relating to several of the preceding Strata, will be given in the subsequent part of the work,

OBSERVATIONS ON ECHINI.

CLASSICAL CHARACTERS.

Roundish, covered with a bony crust, more like a crustaceous than a testaceous animal, with moveable spines, and two apertures, a mouth always on the base of the shell, and a vent variable in its position.

From the different situations of the vent the class is susceptible of three principal divisions :—*Anocysti, Pleurocysti, Catocysti.*

1. *ANOCYSTI.*

Mouth on the base of the shell; vent vertical.

Under this division are comprised two genera, Cidaris, and Clypeus.

CIDARIS.

Gen. Char.—Circular, or ovate, with porous rays, diverging equally on all sides, from the vent to the mouth: vent vertical; central; mouth beneath, central.

CLYPEUS.

Gen. Char.—Circular, or suboval, irregular, with porous or striated rays.

2. *PLEUROCYSTI.*

Mouth on the base of the shell; vent on the side.

SPATANGUS.

Gen. Char.—Cordate or heart-shaped, with porous rays, and a groove or channel from the vertex to one end; mouth beneath, near the broad end, vent at the narrow end above the margin of the base.

Gen. Char.—Oblong, with ten biporous rays in five pairs; vent at one end, above the margin of the base; mouth beneath, almost central.

Echinites lapis cancri is of this Genus.

3. *CATOCYSTI.*

Both apertures on the base of the shell.

CONULUS.

Gen. Char.—Nearly circular, with porous lines from the vertex to the mouth: mouth beneath, central; vent on the circumference of the base.

GALEA.

Gen. Char.—Ovate, with porous rays from the vertex to the mouth, near the broadest end of the base; vent at the narrow end of the base.

The mouth of the Echinus is always in the base of the shell.

Central in Cidaris.

Nearly central in Clypeus.

Nearest to one end in Spatangus.

Excentric in Echinites lapis cancri.

Central in Conulus.

About one sixth of the length from the rounded end in Galea.

Nearly all the species of Echini agree in having ten rays of pores proceeding in five pairs from the vertex of the shell toward the mouth in the base, although they do not in all continue to the mouth (as in Scutum and other Genera).

The rays differ much in the number and arrangement of the pores; in the greater proportion of species, however, the rays are actually double, or each composed of two lines of pores.

The rays in Clypeus sinuatus are of a peculiar construction; they are transversely striated, with a line of pores on the side.

In Scutum, and some other recent genera, the rays make a floriform figure on the surface.

Many species of fossil mammellated Echini have articulating edges surrounding the small spherical tops of the mammellæ, which are generally punctated.

The Stratigraphical Table exhibits many instances of repetition of the same species in more than one stratum, and a little attention evinces the accuracy of an observation previously made, that "*similar strata contain similar fossils*;" Clypeus, No. 2, (clunicularis Llwyd), is a common instance. This species is repeated in five different strata, all of them calcareous, for the fossils in the Clay over the upper Oolite lie within a small depth of the rock, and are generally filled with stony matter.

Yet although the same species is repeated in these different strata, a considerable difference in appearance may be traced between specimens from the Pisolite and others from the Cornbrash and Oolites. These last are thinner at the edge, with a more undulated base and flatter sides, particularly that side containing the groove.

ANOCYS

CIDARIS.

	DESCRIPTIONS AND NAMES.	Chalk	Green Sand	Coral Rag and Pisolite	Cornbrash	Clay over the upper Oolite	Upper Oolite	Fuller's Earth and Rock	Under Oolite
1	Subglobose, circular or pentagonal; rays biporous, depressed; areæ swelling; surface rough with small points. Cidaris granulata. *Leske.*		*						
2	Depressed, upper side convex; apparently without any rows of eminences; areolæ prominent.			*					
3	Circular, depressed; shell thin, rather smooth? Cidaris coronalis? *Klein.*	*							
4	Circular, depressed; ten rows of articulated eminences, two rows in each area.	*							
5	Circular, depressed, rays biporous; twenty rows of articulated eminences, which are nearly equal on the base, but unequal above, the two rows in each areola diminishing faster than those in the areæ; edges of the mouth turned inwards.	*	*						
6	Circular, depressed, with thirty rows of miliæ, two rows in each areola, and four rows in each area; rays biporous.		*						
7	Rather pentagonal, depressed, top convex; rays biporous, two rows of small eminences in each areola, forming the angles, two rows of miliæ in each area; mouth large, its edges turned inwards.		*			*			*
8	Pentangular, depressed, with prominent small mammellæ; two contiguous rows in each areola, forming the angles; four converging rows in each area, the two middle rows short and only on the side; rough with small points encircling the mammellæ; rays obliquely triporous.								*
9	Pentangular, depressed, with thirty rows of small sharp mammellæ, which are almost equal on the sides; the row on each side of the areæ smaller than the others; rays obliquely triporous; mouth ten-sided, with a large notch at each angle.			*					

(STRATA CONTAINING ECHINI.)

The

STRATIGRAPHICAL TABLE OF ECHINI.

TI. | PLEUROCYSTI.

		10	Circular, rather depressed, with ten rows of contiguous mammellæ, (near the vent the first mammellæ in the rows are encircled), two rows in each area, separated by curvilinear rows of small eminences, partly surrounding the mammellæ; areolæ with minute points encircling two rows of eminences; mouth large, decagonal, with a deep notch at every angle: mammellæ punctated, articulated.			*	*	*			
		11	Globose or conical, base flattish; ten rows of very prominent mammellæ, two rows in each area, separated by two rows of small eminences; each side of the area bordered by a row of small distinct eminences; rays subflexuous, biporous; areolæ narrow, edged by two rows of eminences enlarged on the margin, widening from the vent downwards to the margin; vent small; mammellæ not encircled by a ring of points, the small globose top of each punctated, and surrounded by an articulated margin.			*					
		12	Rather high; ten rows of mammellæ, two rows in each area, five mammellæ in a row; areolæ narrow, flexuous; apertures large, mammellæ punctated at top, articulated.		*						
		13	Subglobose, with ten rows of large mammellæ, each a little sunk and surrounded by a ring of small points; two rows in each area, separated by numerous small points; rays double, biporous, subflexuous, enclosing two rows of small points: many distinct points on each side of the area: apertures large. The small globose top of each mammella is punctated, and surrounded by an articulated margin.			*					
		14	Subglobose, with ten rows of distant mammellæ surrounded by seven eminences; two rows in each area; areolæ narrow, with two rows of small eminences; mouth not very large. Mammellæ punctated at top, articulated.		*						
CLYPEUS.		1	Clypeus sinuatus. *Leske.* Circular, depressed, convex above, base rather concave, with five diverging furrows; rays ten, in five pairs, approaching each other on the circumference, transversely striated, with a line of pores on their margin: one of the areæ divided by a deep sharp furrow from the vertex; mouth small, pentagonal; shell covered by small eminences in circular depressions.								*
CLYPEUS.		2	Oblong, subquadrangular; base concave in the middle; mouth small, five-angled; upper side convex, with a large deep furrow on one side from the apex to the margin: rays ten biporous lines in five pairs, depressed on the base: apertures opposite, excentric from the broad or furrowed end. Shell unequally covered with small granula, most numerous on the base.			*	*	*	*		*
SPATANGUS.	RAYS IN FURROWS.	1	Oblong, cordate, rather high, margin rounded; dorsal furrow narrow, the ray contained in it not distinct; dorsal ridge rising higher than the apex: covered with granula, particularly on the middle of the base.	*							
SPATANGUS.	RAYS IN FURROWS.	2	Cordate, depressed, dorsal furrow narrow; covered with granula, particularly on the middle of the base.	*							
SPATANGUS.	RAYS IN FURROWS.	3	Cordate, margin rounded, dorsal furrow large, ridge high; height of the shell two thirds of the length.	*	*						
SPATANGUS.	RAYS IN FURROWS.	4	High, base flattish; rays short; dorsal furrow wide, not deep, ridge rising very high; vent as high as the apex; length one third of an inch, height and breadth one fourth.		*						
SPATANGUS.	RAYS NOT IN FURROWS.	5	Long-cordate, the top rounded, with a short deep groove to the mouth; rays obsolete.	*							
SPATANGUS.	RAYS NOT IN FURROWS.	6	Cordate, or roundish, base flat, top convex or rather flattened, dorsal furrow short; rays ten biporous lines diverging in pairs: vent oblong. Much resembles (if not the same) Spatangus subglobosus. *Leske.*	*	*						

above Colours represent on my Maps and Sections the courses and extetent of the Strata containing E

STRATIGRAPHICAL TABLE OF ECHINI.

...TI.							*PLEUROCYSTI.*					
					CLYPEUS.		SPATANGUS.					
							RAYS IN FURROWS.				*RAYS NOT IN FURROWS.*	
Circular, rather depressed, with ten rows of contiguous mammellæ, (near the vent the first mammellæ in the rows are encircled), two rows in each area, separated by curvilinear rows of small eminences, partly surrounding the mammellæ; areolæ with minute points encircling two rows of eminences; mouth large, decagonal, with a deep notch at every angle: mammellæ punctated, articulated.	Globose or conical, base flattish; ten rows of very prominent mammellæ, two rows in each area, separated by two rows of small eminences; each side of the area bordered by a row of small distinct eminences; rays subflexuous, biporous; areolæ narrow, edged by two rows of eminences enlarged on the margin, widening from the vent downwards to the margin; vent small; mammellæ not encircled by a ring of points, the small globose top of each punctated, and surrounded by an articulated margin.	Rather high; ten rows of mammellæ, two rows in each area, five mammellæ in a row; areolæ narrow, flexuous; apertures large, mammellæ punctated at top, articulated.	Subglobose, with ten rows of large mammellæ, each a little sunk and surrounded by a ring of small points; two rows in each area, separated by numerous small points; rays double, biporous, subflexuous, enclosing two rows of small points; many distinct points on each side of the area: apertures large. The small globose top of each mammella is punctated, and surrounded by an articulated margin.	Subglobose, with ten rows of distant mammellæ surrounded by seven eminences; two rows in each area; areolæ narrow, with two rows of small eminences; mouth not very large. Mammellæ punctated at top, articulated.	Clypeus sinuatus. *Leske.* Circular, depressed, convex above, base rather concave, with five diverging furrows; rays ten, in five pairs, approaching each other on the circumference, transversely striated, with a line of pores on their margin: one of the areæ divided by a deep sharp furrow from the vertex; mouth small, pentagonal; shell covered by small eminences in circular depressions.	Oblong, subquadrangular; base concave in the middle; mouth small, five-angled; upper side convex, with a large deep furrow on one side from the apex to the margin: rays ten biporous lines in five pairs, depressed on the base: apertures opposite, excentric from the broad or furrowed end. Shell unequally covered with small granula, most numerous on the base.	Oblong, cordate, rather high, margin rounded; dorsal furrow narrow, the ray contained in it not distinct; dorsal ridge rising higher than the apex: covered with granula, particularly on the middle of the base.	Cordate, depressed, dorsal furrow narrow; covered with granula, particularly on the middle of the base.	Cordate, margin rounded, dorsal furrow large, ridge high; height of the shell two thirds of the length.	High, base flattish; rays short; dorsal furrow wide, not deep, ridge rising very high; vent as high as the apex; length one third of an inch, height and breadth one fourth.	Long-cordate, the top rounded, with a short deep groove to the mouth; rays obsolete.	Cordate, or roundish, base flat, top convex or rather flattened, dorsal furrow short; rays ten biporous lines diverging in pairs: vent oblong. Much resembles (if not the same) Spatangus subglobosus, *Leske.*
10	11	12	13	14	1	2	1	2	3	4	5	6
							*	*	*		*	*
		*		*					*	*		*
*	*		*			*						
*						*						
*						*						
						*						
					*	*						

above Colours represent on my Maps and Sections the courses and extxtent of the Strata containing I

CATOCYSTI.

Genus	No.	Species	Description								
		Echinites lapis cancri. *Leske.*	Obtusely ovate, gibbous, vertex not central, with four pores; rays subpetaloidal, biporous, distinct almost half the height from the vertex, then suddenly lost, appearing again on the base in a cinque-petaloidal star round the mouth; vent at the widest end over a small furrow, round; mouth opposite the vertex, small, pentagonal: base flattish.		*						
CONULUS.	1	Echin. albogalerus. *Leske.*	Nearly circular, conical, vertex perforated by five holes; rays ten, biporous, in five pairs; mouth small, round, vent oblong; shell covered by small eminences; base flat.	*							
	2	Echin. vulgaris. *Leske.*	Nearly circular, convex or subconical; rays ten, biporous, in five pairs; mouth small; base flattish, rather undulated.	*							
	3		Pentagonal, depressed, rays of pores ten, biporous, flexuous, depressed; the areæ swelling into rounded angles; mouth and vent rather large.	*							
	4		Convexo-conical, circular, base concave in the middle: rays ten, closely biporous, diverging in five pairs, the two rays near together separated by a flat rising; mouth ten-sided, vent oval.		*						
	5		Convex, depressed, nearly circular, covered by small prominences particularly on the base of the shell, rays ten biporous lines diverging in five pairs; the pores very close on the upper part, more distant on the base, which is concave in the middle; apertures large, vent oval.				*			*	*
GALEA.	1	Echin. ovatus. *Leske.*	Ovate, shell distinctly assulated; rays ten, biporous, in five pairs, the pores opposite, closest and most apparent near the vertex; base flat, rising in the middle, with many scattered granula and intermediate minute puncta; vent at the narrow end, roundish; mouth transverse, with several granula near it.	*							
	2	Echin. pustulosus. *Leske.*	Ovate, narrowing at one end; rays pustulose, continuing from the vertex to the mouth, each pustula concealing two pores.	*							
	3		Subelliptical, round topped, not very pointed at the end; rays biporous on the top, pustulose on the base; pustulæ smaller than in Species 2.	*							
	4		Oval, very pointed at one end, rounded at the other; rays biporous one third of the height downwards from the apex, distinct on the base; ridge sharp; apex excentric towards the broad end.	*							
	5		Oval, pointed at one end, high; rays biporous two thirds of the height downwards from the apex; areæ flattened.	*							
	6		High with flattish sides, top rounded or flattened; rays biporous downwards to the margin; height almost equal to the length.	*							
	7		Three fourths elliptical, the base, forming the section, rounded, contracted.	*							
	8		Broad ovate, one end pointed; margin of the base undulated beneath with fourteen projections, the areolæ form five, the ridge one, and there are eight others in the four other areæ; a large depression on the upper part of the shell behind the apex.	*							

Echini.

The variation in the form of fossils, according to the nature of the stratum, is so remarkable, that in almost all the species which occur in different strata sufficient distinction may be perceived in the general appearance, to ascertain from what stratum they were collected.

Thus, where artificial specific differences fail, the natural features of relationship are sufficient for their identification.

Powerful glasses are required to discover the beautiful organization of petrified Echini, the delicate minutia of which is the more remarkable as the shells are now entirely converted into opaque crystals of Carbonate of Lime. The rhomboidal fracture of this carbonate of lime has long been considered characteristic of these shells and of their spines, and is common to no other organized fossils, except the Encrini.

Spines of Echini are generally more rare than the shells, but are so abundant in the Coral Rag as to be considered one of its best identifications.

The little fossil Clypeus being not only the common inhabitant of three rocks which have a general resemblance, but peculiar to the oolitic part of two of them, shows most distinctly the necessity and great utility of attending to their localities and to their peculiar sites in each stratum.

The stratigraphical table of fossil Echini comprises all the species enumerated in this work, and contains enough for the general purpose of identifying the British Strata.

By the table it appears that Echini are not common to more than eight Strata, Chalk, Green Sand, Coral Rag and Pisolite, Cornbrash, Clay over the Upper Oolite, Upper Oolite, Fuller's Earth Rock (rare), and Under Oolite.

Thence also it appears they are most abundant in the Chalk and in the Green Sand beneath it; the Genus Galea, containing eight species, is entirely peculiar to the Chalk.

The order Pleurocysti is found only in the Chalk and Green Sand, most of the species are found in the Chalk: three out of the five species of Conulus are found in the Chalk.

Clypeus sinuatus is characteristic of the Under Oolite, and occurs at many places in its course; a similar species is found rarely in the Coral Rag and Pisolite, but it is more convex, and larger.

Printed in the United States
By Bookmasters